"FORCE OF ORDER AND METHODS ..." AN AMERICAN VIEW INTO THE DUTCH DIRECTED SOCIETY

STUDIES IN SOCIAL LIFE

XIX

EDITOR:

GUNTHER BEYER

"FORCE OF ORDER AND METHODS..." AN AMERICAN VIEW INTO THE DUTCH DIRECTED SOCIETY

by

MAURICE C. BLANKEN

MARTINUS NIJHOFF / THE HAGUE / 1976

ISBN-13: 978-90-247-1849-8 e-ISBN-13: 978-94-010-1393-2
DOI: 10.1007/978-94-010-1393-2

TABLE OF CONTENTS

MAPS AND CHARTS

PREFACE

The Netherlands is an unusual nation in many ways. It is not only that nearly half her land is below sea level. Nor is it that she is one of the world's most crowded lands; her more than 13 million people create a population density of about 1000 per square mile. Nor is it that half her national income is dependent upon world trade. Nor is it that so small a nation could achieve peace and prosperity with so little natural resources. What is most unusual is that the Netherlands has made such a rapid and total adjustment to the demands of modern technological society. In no small measure this was achieved by a deliberate policy of planning, direction, control and development. Its postwar history tells how a determined people under intelligent government leadership rose from a broken economy to a level of economic and social development that places their society among the most modern in the world. The Netherlands is a success story that in some measure has been overlooked by a wider world. This will be an attempt to record her story, touching upon some of the causes and results of this success.

The Netherlands is undoubtedly one of the most planning conscious of modern nations. This is not to say that the Dutch government or its people have any concept comparable to the totality of Soviet Five Year Plans. It might be more accurate to see Dutch planning as similar to America's Tennessee Valley Authority project – direction and economic contribution by the government, participation and cooperation by the people involved. Dutch action was multiple and continuous – a series of TVAs – not a solitary accident. Moreover, Dutch planning represents a more flexible form of social effort than what one usually conceives planning to be. The Dutch government suggests lines of approach and the people concerned react. Frequently interest groups and organizations initiate programs and government plans are always open to changes resulting from public pressures. While there is considerable flexibility within planning arrangements, the government is highly professional in its planning. It remains largely outside the political arena. This is true on all political levels. Planning is thorough, attempts to cover all foreseeable contingencies and delves into all phases of Dutch life. Throughout, planning in the Dutch society remains a form of social control directed towards improving the economic and social living standards of the entire society. It is done with a modicum of fanfare and social friction.

Lately there have been high levels of public concern on ecological issues of air and water pollution, use of space, preservation of natural areas and population control. Government awareness of the need for planning, and public acceptance of this need, arose from the particular historic, geographic and economic conditions of Dutch life. Planning is not new to the Netherlands. The building and maintenance of dykes and drainage systems has an ancient lineage. Economic regulation of crafts and trade go back far in her history. Housing regulations and social legislations came to the Netherlands before many western nations. Numerous lessons of restraint and cooperation were learned in the process. However, the more significant, and now increasingly total direction of Dutch society, is new. Post World War Two is the watershed. The inter-war depressions of the thirties, and the harshness of the Nazi Occupation almost destroyed Dutch economic life. These events also broke up the social order and local parochialism so indigenous to the pre-war Holland. In its place there was created a wider feeling of nationalism and a heightened popular insistence upon greater social equality. Neither the government nor the people could ignore their economic plight. Their amazing post-war recovery, although the Dutch are extremely grateful, cannot be totally attributed to Marshall Plan Aid. Their economic successes whetted their optimism and gave the Dutch a forward looking attitude which supports the planning and direction for social and economic development still underway in Dutch society. A new Netherlands has been born. She is now a complex industrial, technological society – more a wave of the future than a sleepy backwater of windmills and wooden shoes.

Growing dilemmas – perhaps not fully apparent to the Dutch people and their leaders – are developing from these changes. The Netherlands is, for the most part, enjoying a booming economy that undergirds her hopes for continued economic and social development. Still her economy is geared to the exigencies of the world market – or at least the economy of Western Europe. Willy-nilly, the Netherlands is tied to the vagaries of world trade. Her home market cannot sustain her domestic labor force who come seeking jobs. It is saddening to contemplate how greatly her people may suffer in a world depression. Another equally critical dilemma arises also from her post-war boom. Economic well-being has encouraged a population growth that was a decade ago the highest in Europe. The Netherlands has fought the sea, has wrested from it more land for her people, and held back the ever recurring threat of floods. The new floods she must dread will be the sea of people that will strain her land resources to the bursting point. What dykes and Delta Projects are there for this threat? These two dilemmas are parents to the third. Increased industry and population are making exigent demands upon the physical environment of the Netherlands so that ultimately ecological collapse may ensue. Already land, fresh water, natural beauty, open space, quiet privacy and "breathing room" for sane living are at a premium. Seemingly industrial and population growth can persist only if the technological replaces the natural environment. Perhaps the Dutch are prepared for this eventuality more than people in any other society. Still there is always a lurking vulnerability in technological complexity. Is the Netherlander any more responsible or capable than any one else of maintaining the degree of sophistication and self-control that a technological society demands?

Recently, as throughout the world, events in the Netherlands have moved rapidly. What began in this study as social comment seems already to have become history. The former comfortable stability of Dutch Society appears to be confronting the greatest challenge to its traditions and institutions. The capacity of her advanced technology to continuously solve her economic, social and ecological dilemmas seems overstrained. In the long run more may depend upon the character and social philosophy of her citizenry than upon 20th century technology. Despite the best of intentions, and the kind assistance of both the editor and publisher, which the author now wishes to acknowledge, it has been virtually impossible to keep abreast of the attempts the Dutch have made to cope with their problems. Regretfully, by the time of publication, this study may inevitably contain some elements of outdated fact which may affect its findings. For this the author alone takes responsibility.

The Dutch are a fine people. They are sincere, honest, industrious and in no way malicious. Their dogged industry and unfailing courage in overcoming the challenges of their past, their general goodwill to mankind in the present, ought to be rewarded by good wishes in return. Their future is important to us all. Since the Netherlands is the most demographically concentrated society, has the highest mean level of technological development, and is the most planned and directed society, there is a high probability that her fate portends mankind's future.

CHAPTER I

THE DUTCH "MIRACLE"

Increasingly the critical problems of the late 20th century are becoming ecological in nature. Economic geography may well be the most determining factor in a nation's well-being. Perhaps no modern nation has so boldly and thoroughly confronted her ecological limitations as has Holland. Here there is no prodigality of land that one finds in Brazil, Canada, America or the Soviet Union. Holland has barely enough land for a livelihood, homes and recreation. What is available is largely due to the diligent uses to which the Dutch have put their land. All must be carefully husbanded and judicious decisions must be made continuously concerning the use of their land. This the Dutch have largely done and up to now have avoided the disaster that would surely have been theirs had they employed the waste and wrack of the more generously endowed nations. There are indications that her present affluence has created trends – the wider personal use of the automobile, the spread of suburbia outside large urban areas, and the more wasteful, litter producing packaged-goods consumption – that might be wiser to avoid. However, rather than lay waste their land, they have ever attempted to improve it. Perhaps no better illustration exists to show the impact of geographical endowment upon social attitudes and behaviour than the fact that a geographically poor Holland, during its existence, increased its arable land by 50 percent, while a richer America, through ill use and erosion, decreased hers by an equal amount in less than 200 years.

Nature has not been overkind to Holland. Even were we to discount her climate the statement would remain true. Perhaps nature's most generous gift was the alluvial soil of her neighbors that made Holland a possibility. When Napoleon quipped that God made the world, whereas the Dutch made Holland, he failed to point out that in some measure it was at the expense of the French whose soil helped fill

Dutch marshes. In the geological past, restrained by off-shore islands and their sand dunes, the rivers of north-western Europe deposited their burdens, gradually silting in that area now called the Netherlands. Marshes and bogs developed in this low region behind the natural sealine. So thorough was this silting that at the time of the Roman occupation the Zuiderzee was enclosed and known as Lacus Flavus. Now this process is reversing itself as the sea eats into the former marshes. Two penetrating sea-arms, one down from the north through the Zuiderzee, the other up from the south through the Schelde, have been threatening the lowlands for centuries. Today the IJsselmeer and Delta projects are Holland's answer to the sea's challenge. In ages past there were the dunes and the dykes to stem the tides. Nearly half of Holland's land came into being only through the industry of the Dutch in the construction of their polder land. Thus even her land is not fully a gift of nature.

Nor had nature been generous to Holland in known natural resources. The former marshes did yield some peat and there are coal resources in the Dutch "Highlands" of Limburg. Properly treated, the soil of the Netherlands is favorable to agriculture. It remains the world's highest per capita consumer of chemical fertilizers. This niggardliness of nature has been greatly ameliorated by the 20th century. In the twenties large pits of salt were found in the Northeast. During the fifties one of the greatest fields of natural gas in all Europe was discovered in the same region. This resource has added substantially to the nation's income. Having previously shifted from oil to gas heat, the nation averted the more inconvenient aspects of the Arabian oil embargo imposed upon her due to her friendly attitude towards Israel. There are now known reserves of oil and natural gas off the country's northern coast, which promise to be profitable. While somewhat disappointing to the Dutch, in comparison to the gigantic oil finds off the coast of Norway, the good fortune of her neighbor will compensate somewhat by bringing indirect benefits to the technical, construction, shipping and warehousing interests of Holland, not to mention a cheaper and more assured source of oil.

Holland, in her geographic site, is certainly most blessed. She is located almost in the heart of the population center of Western Europe – London and environs to the northwest, Antwerp-Brussels to the south, and the Ruhr to the southeast. More population and markets lie beyond. There are over one hundred and fifty million people within a

375 mile radius of the Amsterdam-Rotterdam axis. The important river systems of Northwest Europe – and their connecting canalways – converge upon the lowlands. In addition, Holland has excellent sea outlets and harbors. The combination of these factors makes Holland the entrepôt of Europe.

The Dutch have capitalized upon this beneficial accident of nature and under government direction have created one of the most modern, efficient and profitable port, shipping, warehousing, transport, communications and banking networks in the world. Rotterdam is the busiest of world harbors. The Dutch, not content with this, are erecting extensive port facilities called Europoort, out into the sea. They have already built one of the world's largest petroleum complexes along Rotterdam's waterway. A vast chemical industry has been cultivated alongside this petroleum industry, and plans were under discussion to construct the nation's largest steel plant here also. As an entrepôt and industrial center, the Rijnmond will have no equal in Europe – possibly in the world. The very success of Rotterdam as a port and industrial center makes it so densely compact and congested that it is becoming an undesirable living area. While it is the heart of the largest urban region of Holland, it is also possibly the least pleasant. The problems of space and air and water pollution are so intensified that authorities are placing heavy restrictions upon existing and future industries.

Another center of shipping and industry converges upon the IJmond. Amsterdam, Holland's second port, maintains its life only because the Dutch have built the IJ and the world's largest shiplocks at IJmuiden. Three times, in less than a century, the Dutch have rebuilt these locks to insure Amsterdam's standing in world trade. Yet inevitably such industry will defeat itself. Already Amsterdam's shipping, chemical, petroleum and steel industry settled along the IJmond is strangling for space. Amsterdam's future development will probably depend upon her banking and commercial services, and her light industries.

With characteristic energy, the Dutch are building a third shipping and industrial center at Delfzijl, a port for the gas and aluminum industries that have been developed in this region during recent years. The Dutch are also considering building a port and industrial complex (The Sloe-Project) along the Schelde, for with the completion of the Delta Project they anticipate increased industrial opportunities in this area. Undoubtedly this project will complement, if not rival, the port of Antwerp.

The rivers Maas, Waal and Schelde make Holland the doorway to northwest Europe and the Dutch the traffic bearers of European trade. Dutch carriers are responsible for 27 percent of Common Market shipping and 53 percent of Rhine river traffic. This last is due to the excellent network of canals crisscrossing the land. Transportation to Belgium, France, the Federal Republic of West Germany, and beyond is thereby extremely economical. The Dutch are most careful in the maintenance and regulation of this traffic. Nor have they ignored motor transportation. Dutch carriers account for 30 percent of the commercial road traffic of Western Europe. Despite the problems of frequent water crossings, poor road beds of the low soggy soil, and the resultant exorbitant costs of road construction, Holland's road network is superior to those of Europe. Their superhighways (Autowegen) are narrow but safe. Prior to the oil embargo ban of 65 miles per hour, cruising speeds up to 70 to 80 miles per hour were permitted. The country has the most densely concentrated road system in Europe. Unfortunately, despite frantic efforts, road needs lagged behind constantly rising passenger vehicle registrations. This may correct itself if the Dutch, in a reaction to the "Oil Crisis," move from a reliance upon the auto to a return to the use of public transportation. Holland's two major airports are run as modern efficient air terminals. Schiphol is one of the busiest airports in Europe. Zestienhoven, outside Rotterdam, is used primarily for local, short distance, international travel and commercial traffic.

These recent developments, particularly in industry, have altered the traditional Dutch economy, which rested upon a triad of agriculture, banking, and commerce. Just as present economic success is in a measure attributable to government planning and management, so has it been in the past. Traditionally her leaders have practiced a pragmatic management of the nation's economic and social development. Nowhere was this more evident than in Holland's glorious "Golden Century" when direction of agriculture, investment and commerce was the order of the day.

In those earlier days the Dutch were diligent and productive farmers. They were early breeders of cattle and advanced systematic farming methods long before the more touted efforts of the English. The genre painting of the "Golden Century" clearly shows the level of wealth among the Dutch farming folk of that day.

During the same period, the Netherlands rose into prominence as the banker of Europe. By 1600 Amsterdam was the foremost bourse of the continent. Dutch merchants early recognized the economic limitations of their land and sought economic opportunity and security by shrewd and thrifty investments. They created the Bank of Amsterdam and developed insurance institutions on sound actuarial principles. Her bankers made substantial loans to the Kings of England and France. Holland was the only creditor of the government of the infant American nation up to 1789.

Since the 17th century Holland has been known as the trader of the world. At the beginning of that century she had a trading fleet of 20,000 vessels – more than all the rest of Europe together! This prominence grew out of Holland's near monopoly of the herring and salt trade. Early in the 1600's the Dutch Herring Fleet had grown to such prominence that it was centrally controlled in its seasonal exploitation of the fishing off Scotland, Northumbria and the Dogger Banks. Under the aegis of the College of Fishery, 32,000 fishermen of the "Grote Visserij" (Great Fishing Fleet) were controlled in their fishing and protected by a navy provided for by the Estates General; and wisely so, for the revenue from the "Grote Visserij" by the mid-century was three million pounds sterling – greater than England's cloth industry at that time. Limited ports of entry served as an additional control over this important activity. Scheveningen, one such port, has a coat of arms of three herrings on a shield, attesting to the importance of this industry. The profitable wool trade with Scotland was severely regulated with the town of Veere, the only Dutch port open to this monopoly. The Netherlands dominated the Baltic trade during this period. Many of her Zuiderzee and river towns were thriving members of the Hanseatic League. With the acquisition of her East Indies holdings, she later had a near monopoly of the spice trade. The Dutch East and West India companies were models of mercantilism quickly copied, but with less success by the English and French. By the middle of the 17th century her fleet of 15,000 comprised sixty percent of Europe's tonnage. Gradually, inroads were made upon Holland's shipping monopoly, but as late as 1700 she still had a fleet twice that of England and tenfold that of France.

It was during this period that there was considerable investment in impolderization. Her renowned shipbuilding and diamond industries were also established at this time. It was not mere chance that Peter the Great, who wished to modernize his country, came to Holland to learn how. There, in addition to working in the shipyards, he visited flour mills, paper mills, attended lectures on surgery and discovered the intricacies of the printing press – all the advanced innovations of the age. Then, as now, the Dutch understood their role in the world. They sensibly turned their attention to the production of quality goods with high returns, leaving the lesser profitable trade to others. Their merchants understood well their important role as middlemen. Holland was even then the entrepôt nation of the continent.

> The Dutch must be understood as they really are, the Middle Persons in Trade, the Factors and the Brokers of Europe ... they buy and sell again, take in and send out, and the greatest Part of their Commerce consists of being supplied from All Parts of the World, that they may supply All the World Again.[1]

It was readily recognized throughout Europe that the leaders of the Netherlands understood the art of directing and developing an economy of a nation. Holland's past economic development was no more due to chance then anymore than it is now.

> ... but the result of a great concurrence of circumstances, a long course of time, force of Order and Methods, which never before met in the world to such a degree or with so prodigious a success, and perhaps (we) will never see again.[2]

Organization, planning, direction – these have a long standing and honored position in the Dutch scheme of things. It is little wonder that Louis XIV epitomized the Holland of his day as "le nid de la richesse."

Holland's enviable monopoly of banking and trade and her leading position in agriculture could not long persist. Such success invites competition. It was England that first surpassed the Dutch in banking and trade. The shift in power came after a series of naval wars in which English superiority of numbers prevailed. One of the consequences of the shift, little marked in Holland, but of great importance to the future American nation, was the loss of Dutch holdings in the New World, when in 1664 New Amsterdam became New York. Wars with England were succeeded by wars against Louis XIV of France. The Dutch

[1] Daniel Defoe, *A Plan of the English Commerce*, 1728, quoted in *The Dutch Republic*, Chas. H. Wildon, NY, (McGraw Hill) 1968, p. 22.

[2] Such at least was the opinion of Sir William Temple, England's Ambassador to the Low Countries in 1673. Quoted *Dredge, Drain Reclaim, The Art of a Nation*, J. Van Veen, The Hague, 1962, (M. Nijhoff) p. 83.

stoutly held their own, even in the one war in which she faced a coalition of France, England, Sweden and several German States. A return to peace brought a period of prosperity and complacency to Holland. This was shattered when the uproar of the French Revolution broke in upon her. After a brief spell as the Batavian Republic she fell a victim to Napoleonic imperialism, becoming first a satellite kingdom and eventually for a few years a part of France itself. Holland suffered greatly under the chafing restrictions of the Continental System. The defeat of Napoleon at Waterloo, to which a Dutch army contributed, saw Holland free once more. A result of these imperialistic and mercantilistic conflicts was an appreciation by the Great Powers, particularly England, that the neutrality of the Low Countries was a key to the peace of Europe. Holland's unhappy marriage to Belgium, urged by England, was ended in 1839 and Holland was once again free to tend wholly to her own affairs.

With this external entanglement brought to a close, Holland embarked upon a period of great commercial and industrial growth. Much of this was due to the encouragement and efforts of the government under the restored House of Orange. The rulers were not above concerning themselves directly with the commerce and industry of the country. William the First won the epithet, "The Merchant King" and his citizens considered it an honor when conferring it upon him. These efforts were rewarded for ...

> Dutch commerce yielded large profits, much more so than English, let alone French Commerce. Canals and roads were constructed with great energy, equally distributed over North and South. An efficient spread of the sources of prosperity over smaller centers was planned and implemented. A large number of harbor projects were put in hand. The Netherlands was at that time the second maritime power in the world ...[3]

Some of these government projects assumed giant dimensions. To save the port of Amsterdam whose older water approaches were silting up beyond remedy, the government undertook the construction of the 40 mile long North Sea Canal to Den Helder. This proved to be inadequate almost upon completion, and the government immediately embarked upon the project of the IJ canal and its canal lock system. This was – and after two subsequent enlargement efforts – remains the

[3] "History of the Netherlands," reprint *Digest of the Kingdom of the Netherlands*, p. 18.

largest shiplock in the world. A similar problem confronting the Rotterdam port was solved with the construction of the Calland Canal just before 1870. The government used its resources to encourage investment in industry, the opening of mines, the building of railroads, and the establishment of a large textile industry in the economically depressed eastern region. Many of these efforts served to ameliorate the poverty and economic dislocations occasioned by the Industrial Revolution. The social effects of these industrial changes were no less unpleasant in Holland than in other countries undergoing a similar process. It was during this period that the government turned, somewhat belatedly, to social legislation. The first child labor law was passed in 1874; the first social insurance and the first housing acts in 1901. Political reform under the Constitution of 1848 established a more popularly representative government. Paralleling historic trends throughout Western Europe, Dutch political leadership lacked a liberal cast until the last quarter of the 19th century. Dutch political parties, dominated by religious issues, and their followers who were accustomed to government patriarchal benevolence for their welfare, ignored issues that would have been more responsive to popular needs. Dutch society had not fully weened itself from an attitude of expectancy of government direction and interests in its affairs.

The 20th century began on a happy note. Prosperous years came to the Netherlands. But it was not to last. The advent of 1914 put a severe crimp in the Dutch economy. Isolated as a result of the war, Holland was once again reminded of her precarious reliance upon world markets. Unemployment rose as her economic life fell stagnant. She had barely recovered when the impact of the World Depression demolished the beginning prosperity her shipping, trading and banking enterprises were building for her. If her economy was hurt by the advent of the Depression, it was shattered by World War II German occupation.

No occupied nation of the West was so heavily harmed by Nazi occupation as was Holland. Property loss and destruction came to almost $ 2000.00 per person – an enormous loss considering per capita income after recovery in 1952 was a mere $ 600.00. It was a quarter of a century after the war's end before her *per capita* income reached the level of wartime loss. Thirty percent of Dutch industry was destroyed and forty percent of her productive capacity was lost. Her transportation system was at a standstill. Her railroads had been looted of all but 20 percent of their rolling stock. Almost her entire international fleet had been lost. Ten percent of her housing had been destroyed, and virtually no housing construction occurred during the occupation, this despite a population growth of 5 percent during that period. She had been looted of 20 percent of her cattle, 50 percent of her swine and her poultry industry was almost completely destroyed. The port of Rotterdam, the most wantonly bombed allied city of World War II, was almost totally obliterated. Other towns – Arnhem, Middelburg, Rhenen – were also badly damaged. There was in addition the flooding of Walcheren and Wieringermeer and Northeast polders to reckon with. On top of this, Holland also lost her East Indies colonial holding that contributed so greatly to her trade and processing industries. Her recovery was hampered by several severe winters, and the disastrous floods of 1953 that inundated over 5 percent of the land, killing 1800 and causing great loss of homes, farmland and villages. The five years of virtual economic stagnation in agriculture, industry, trade, shipping, banking and housing made the post-war task of reconstruction that much more formidable. There was a heavy press of growing population demanding first food, then jobs, housing, and better living standards. The Dutch government and her people acted vigorously to meet the challenge. Even in exile and in the underground there were organizations formed to prepare for rebuilding the Homeland. There was a blessing in her adversity. Her problems seemed so infinite that only an all-out effort could begin to encompass them. Holland therefore ventured upon a near totally planned and controlled economic program, that created, in effect, an economic revolution for her society.

The first post-war measures were of an emergency nature dealing with immediate food supplies to avert starvation in the most populated areas, the draining of the flooded lands, restocking the farms, tooling up dormant industries, rebuilding shipping and restoring Dutch trade. This was done under an emergency government program of deficit spending. Foreign aid was extremely helpful, above all the Marshall Plan Aid. Holland received $ 1.1 billion in ECA aid beginning in 1948. So rapid was her recovery that the Dutch government was able to announce her departure from the Marshall Plan program by 1953. Her national income had risen from $ 3.9 billion to $ 5.6 billion in a brief four year span! The Marshall Plan spared Holland the necessity of liquidating all Dutch assets abroad, assets vitally important for the maintenance and expansion of her world trade network. Without Marshall Plan Aid, consumption, in an already belt-tightened economy, would have been reduced by five percent. More important would have been the loss of over 25 percent of capital investment in the five year period. Consumption is basically inelastic, and with so little available for saving and investment, it is doubtful if the Dutch economy could have gotten itself off dead-center without foreign help. Marshall Plan Aid proved the sparkplug of economic development in Holland as it proved elsewhere in Europe.

Under government planning, a second stage of development aimed at economic growth was embarked upon. Government goals were full employment, balanced trade and an increasingly equitable distribution of the national income. This was to be achieved by growth of industry with government planning and support. The government was also to be vigilant in protecting the Dutch market and earning power by the careful regulation of wages and prices. High wages meant the inability of Dutch industrial and agricultural exports to compete favorably in the world market. High prices meant the erosion of earning rewards at home. As her economic situation improved, greater rewards in wages, housing and social welfare benefits could be granted the worker.

The government planned increased industrial investment, offering a favorable climate of tax benefits and labor discipline. This last came as a result of the intelligent cooperation of labor and employee organizations with government goals. Though wages were kept to a minimum, there was little strife up to the end of the sixties. The government has upheld its part of the bargain and granted benefits whenever it felt it could without endangering its policies or the national interest. There was a planned shift of investment and labor from the agricultural and trade to the industrial and service sectors. Every effort was made to advance productivity levels. Holland's industrial production which had doubled from 1920 to 1940 was more than halved before the end of the Nazi Occupation. Before ten years elapsed following the liberation, her industrial production trebled, surpassing 1920 levels by 50 percent. This was accompanied by a somewhat matching performance in industrial productivity per worker, which doubled during the same period. Recovery following difficult years of depressions or military occupation can be considered normal, but it is the sustained nature of her economic growth up to now which converts Holland's post-war recovery into a "miracle." "In the 1950–60 decade the real growth of Gross National Product averaged 5 percent annually, or higher than almost any other European country, and during the same period, the average unemployment was only 1.9 percent of the labor force."[4] The same conditions prevailed through the decade of the '60's. National income, which was $ 4 billion in 1945, rose to $ 38 billion in 1969 to become $ 65 billion in 1974. It is this buoyant prosperity that Holland's planners hope to use to underwrite the costs of the future economic and social development of their nation. However, as recent events have made clear, the economy of the Netherlands is highly dependent upon that of western Europe and the rest of the world. The Dutch have once again been reminded that their struggle for economic and social stability has yet to be determined.

[4] *Organization for Economic Co-operation and Development*, "Netherlands" Paris, 1963, p. 27.

CHAPTER II

THE INSTRUMENTS OF DECISION

The way in which the Dutch have chosen to rule themselves may not seem as democratic to the American as it does to them. Their system is neither as simple nor as egalitarian as our own. In the past there have been heavy elements of paternalism in Dutch government. A Calvinist middle class sense of responsibility runs as a thread through the warp of Dutch history. Coupled with a Catholic sense of social conscience, the modern result has been to create a centralized government dedicated to the economic and social well-being of its people. Competition and rampant self-interest is no longer the Dutch way. The populace now expects and trusts its government to manage the country's affairs for their benefit.

A small country with a large population density with limited resources and a heavy dependence upon world trade, the Netherlands can hardly afford freedom from controls. Since it is a small country, political power and political activity naturally gravitated to the central government which today directly and thoroughly affects the lives of every citizen. While intelligently accepting them, the populace will permit no more government controls than it deems necessary. Controls tend to relate to economic and social needs. Few, if any, apply to culture, values or thought. The Dutch are zealous in protecting their freedoms – property, movement, speech, assembly, education, and above all, religion. Their tolerance of differences and dissent shows a more sophisticated appreciation of freedom than Americans are apt to demonstrate. Though a constitutional monarchy, a unitary, centralized state, and a directed economy, the Netherlands remains one of the world's outstanding democracies.

The government of the Netherlands is based on the Constitution of 1848 as amended, with the latest modernization in 1963. Further modernization is expected in the near future. It possesses the traditional branches of government but no built-in system of checks and balances. The Dutch Government arrangement corresponds closer to the British system than our own. Like England, there is a Crown, Cabinet government, Parliamentary democracy, and party responsibility. The Dutch Crown possibly has more power in government affairs than its British counterpart. The Crown is responsible for forming new Cabinets. Little legislation that conflicts with royal views manages to get passed. The Crown also serves as the link with the former Dutch colonial holdings as the Dutch Antilles, that are now integrated into the Dutch Kingdom. The Royal family also plays an important if informal role in foreign relations and domestic government oversight. The House of Orange is held in considerable esteem and affection by the Dutch people. Frequently, appeals for help or justice are made directly to the Queen – and are invariably answered. The Queen serves as a symbol of national unity and adds an element of color in a down-to-earth, matter-of-fact society. However, as in England and other countries with a constitutional monarchy, there is an increasing disinterest in the concerns of the Royal House, particularly among the younger generation.

Two governing bodies are connected with the Crown, that are not found in our own government. These are the Council of State and the General Auditing Court and they help to maintain the integrity of the State and responsibility in fiscal matters. Members of both bodies are royally appointed, and hold tenure for life on fixed salaries, similar to our Supreme Court justices. Thus outside the stream of politics, they lend prestige and objectivity to government and the Crown. The twenty member Council of State is a gathering of "elder Statesmen" responsible only to the Crown. It reviews all new laws submitted to Parliament, all proposed international agreements, and serves as a final court of appeals where government administrative issues can be resolved. It also offers advice to the Crown. The General Auditing Court is similar to the Office of Comptroller General found in many states of the Union and in our larger cities. Operating a post-audit system, it checks upon income and disbursements, holding the Cabinet and government agencies accountable for the handling of public monies. Independent, objective, highly professional, it keeps those involved with public finances to norms that we might consider outstanding levels of fiscal honesty. The Court, with its small 200-man civil service staff, represents a high level of competence and expertise. Management of this professional sort is highly regarded by the Dutch. They appear to prefer it to the political, whenever the government function seems technical enough to make it feasible. However, there has been, none the less, growing criticism of the court by the public and Parliament of what is deemed as arbitrary and uneconomical use of government funds.

The responsibility for control and direction of the government rests with the Council of Ministers. Informally this is called the Cabinet, or interchangeably, the Ministry or the Government. A new cabinet must be formed after each four-year election. The Cabinet is formed under the direction and approval of the Crown. It is given responsibility for the creation and execution of government policy. It is invariably a result of a coalition of the leading Parties seated in the Lower Chamber. Therefore, the Cabinet, like that of the British system, can usually count on legislative support for its measures or policies unless a parliamentary crisis arises. Since cabinets, or governments, are normally the result of coalition, there is considerable political bargaining beforehand and policies tend to be careful and conservative. Many of the issues ride on budgetary, trade policies, wage and full-employment controversies. The present cabinet consists of approximately 33 political posts – 16 State Ministers and 17 Undersecretaries (Staatssecretarissen). The Cabinet positions reveal those areas that are important in modern Dutch government. The Minister-President heads the Department of General Affairs. Recently there has been an increase in the power held by the Minister-President, so that it would not be incorrect to view him as a sort of Prime Minister. There are the traditional posts of Finance, Justice, Foreign Affairs, Domestic Affairs and Defense. Three departments are concerned with the domestic economy – Economic Affairs, Agriculture and Fisheries, and Traffic and Water Control. Five Cabinet posts concern themselves with domestic social needs – Social Affairs, Public Health, Housing and Physical Planning, Education and Science, and Culture, Recreation and Social Work. Two Ministers without Portfolio have been newly created; one for International Development Aid and one for Scientific Advancement and Science Education.

A comparison with our own government administrative organization shows much similarity in the traditional functions of Finance, Justice, Foreign Affairs, and Defense. The Department of General Affairs corresponds to the President's White House Staff. While not identical, the Dutch departments of Economic Affairs, Traffc and Water Control, Agriculture and Fisheries are counterparts to our depart-

ORGANIZATION CHART OF THE GOVERNMENT OF THE NETHERLANDS

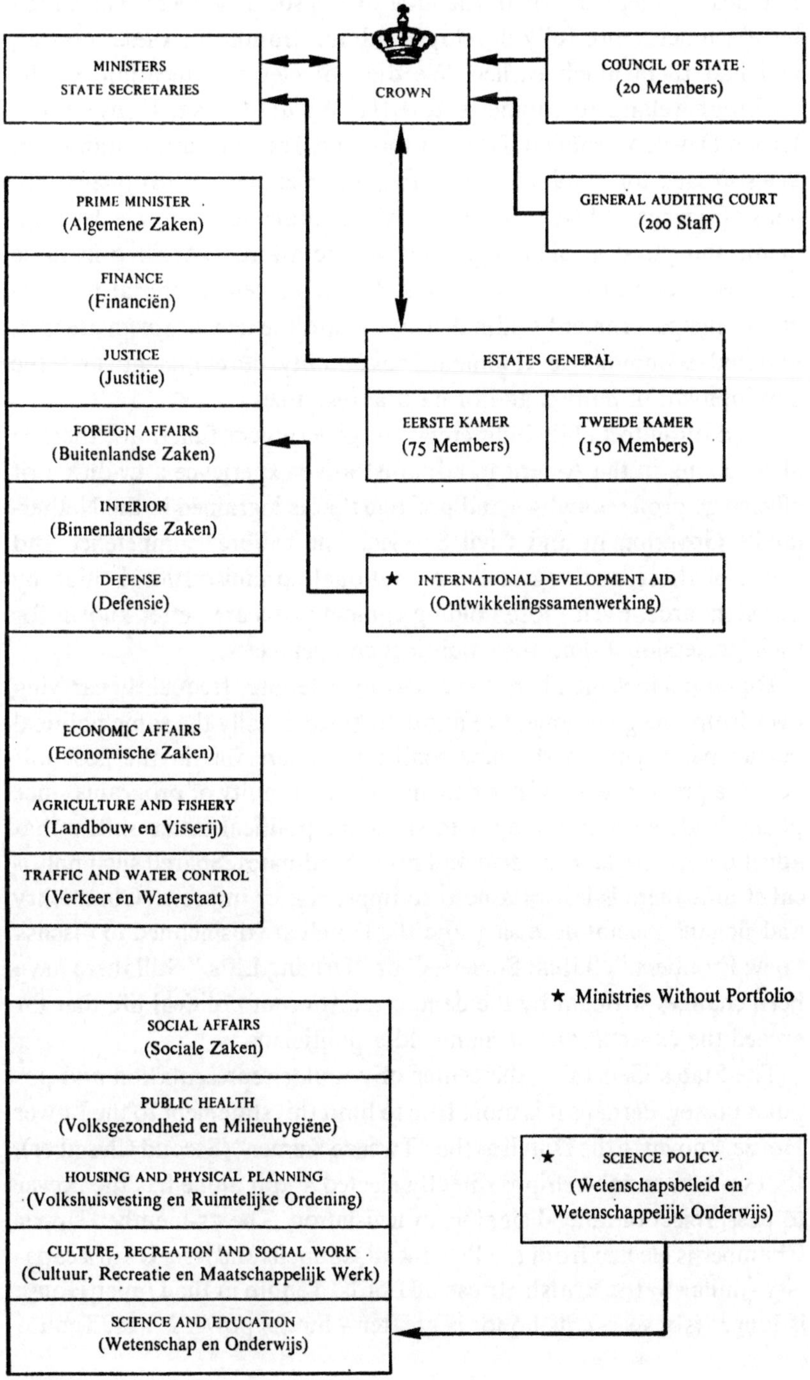

ments of Commerce and Labor, Transportation and Agriculture. The significant differences lie in the area of the social services. The Dutch have a much more fully developed administration for these services, and had them much earlier. We did not elevate education, public health or welfare to cabinet status (HEW) until 1953, Housing and Urban Development (HUD) until 1965 and Transportation until 1967. We still lack top level administrative organization for national problems of science and technology, social, cultural and economic development, and physical planning. That we are aware of trailing in these areas is attested to by ex-President Nixon's recent proposal to reorganize the government's administration into the new problem areas of national economic development, community development, and the development of human and of natural resources.

In addition to public indifference to government functions, there is also lacking in the American administrative experience a tradition of efficiency, professionalism and prestige that is ingrained in the Netherland's Government and Civil Service. The calibre, competence and status of those working in Dutch national administration is high by our standards. Even those holding cabinet posts are better known for their professional than their political competencies.

Dutch cabinet members have had long tenure, frequently carrying over from one government to another, since usually the same political parties participate in the new coalitions. There was in the post-war period a greater sense of permanence and continuity of programs since policy leaders are not subject to the same political uncertainties that affect the American president and his subordinates. Spared such political strains there is less of a need to improvise or impress. Gimmickry and slogans are not necessary and the Dutch are disinclined to discuss "new Frontiers," "Great Societies" or "Driving Lifts." Still there have been changes wrought by the democratization of political life that far exceed the expectations of many elder politicians.

The States General is the center of popular representation and popular power. Perhaps it is more true to limit this statement to the Lower House, known to the Dutch as the "Tweede Kamer" (Second Chamber). Its 150-man membership is directly elected and it alone has the power to pass, reject or amend significant legislation. The 75-member Upper Chamber is elected from the Provincial Councils and it serves in a capacity similar to the British House of Lords. Tenure in the Upper House is longer (six years), its honor is greater – but its power is less. Tenure in the Lower House is four years, unless Parliament is dissolved. Its members are entitled to expenses, allowances and pensions. The "Tweede Kamer" hammers out all the laws supporting the policies of the Cabinet, or Ministry. Since Cabinets are usually formed from party coalitions representing Parliament majorities, there is normally a high level of cooperation by Parliament to effect Cabinet measures and proposals. However, Ministry bills can be amended or rejected. Parliament committees can carry out investigations somewhat like our own Congressional Committees. Moreover, similar to the British system, Parliament can, through its powers to question, or interpellation, considerably embarrass the government in measures or policies with which it feels uneasy. There has been little challenge by Parliament of Cabinet, Ministry, or Government bills. The kinds of decisions that come before the government are more and more technical in nature. Water and traffic control, import and export levels, employment, housing needs, social services, and environmental controls, are increasingly technical and areas for the professional. They are not controversial in nature, particularly since the Dutch seem to have arrived at a consensus as to what they anticipate from government – high living standards, equitable distribution of national income, and more pleasant and liveable environments. In the past, a visit to the "Tweede Kamer" was not a particularly exciting experience. A question as to whether to declare a certain national holiday might have engendered a more heated and lengthy debate than the annual appropriations. Now with the growing intensity of political differences, the uncertain economy and the sharpening of views as to the best technical answers for the nation's social, economic and ecological future, public involvement has increased. Party activity, over-crowding of the tribunes, and demonstrations in front of the Parliament at the Binnenhof, has made Dutch politics, if more confusing, certainly more stimulating.

Party politics in the Netherlands would seem incomprehensible to Americans. Accustomed as we are to a two-party system and the avoidance of religion in politics, the multiple, confessional-party structure of the Netherlands would disturb us greatly. Only as we understand, as the Dutch do, that religion is largely irrelevant to most of government issues, would we be able to accept their politics. Three of the Dutch political parties important to the present government coalition, though in diminishing degrees, maintain ideologies that bespeak religious world views. While evidently such views are fading, it will probably be with no greater rapidity than the demise of our own political irrelevancies such as the Cold War or Republican-Democratic "Tweedledee-Tweedledum-ery."

Religion is significant to the Dutch, and much of the history of the Low Countries has been related to religion. The Dutch Republic came into being out of a struggle that was as much religious as it was national. Their rebellion was against religious oppression as well as political tyranny. It is not without significance that the Dutch National Anthem is a hymn. When Calvinism triumphed, it initially proclaimed for itself the role of a national religion. Carried away by the successful revolt against the might of Spain, and the unexpected, unprecedented wealth of the Golden Century, there was an understandable tendency towards arrogance and fanaticism. The more liberal, humanistic elements of society, represented by Dutchmen of the middle class, burgher stamp, made attempts at moderation. The remaining Catholics were reduced to second class citizenry. Much of Dutch internal history – the downfall of Grand Pensionary Oldenbarnevelt, the exile of Hugo de Groot, the murders of the De Witt brothers – can be explained by the interplay of these religious and ideological differences. Even Dutch reaction to the liberalism of the French Revolution – an "either-or" acceptance or rejection – took on the aura of religious fervor. At the time of the 1848 Constitutional reform, new party alignments, still suggestive of old religious differences, came into being. The main political divisions were between Conservatives and Liberals with the issue of Catholicism versus Protestantism cutting across to make up the four-way "pillarization" (verzuiling) that exists in party commitments to this day.

In 1848, the Conservatives wanted to keep the authority and power of the House of Orange in the ascendency. The king, Willem II, had mistakenly granted a new constitution fearing a popular uprising. It was an over-reaction to the clamour of the poor who were in despair but hardly in rebellion. While 16 to 20 percent of the population were on poor relief, few contemplated overthrow of the order. But the cries for reform placed the Liberals in power and they wished to widen democratic and social reforms. The Liberals were secular and humanistic, reflecting the influence of the French Revolution and the new undercurrents of the rising socialist mood. Their basic middle class membership received, and gave, support to the working-class poor. The Conservatives' support came largely from staunch Calvinist Protestants.

Its members were anti-Catholic and anti-revolutionary. Catholicism continued to suffer economic, social and political disabilities. None of these were true political parties.

The first movement to be more political than ideological was the Anti-Revolutionary Party (ARP) founded under the aegis of G. Groen van Prinsterer in 1878. These were the Protestant Calvinist supporters of the Conservative Movement of earlier times. They were hostile to the liberal and secular doctrines of the French Revolution. They were opposed to the social and labor legislation engineered by Thorbecke, "the Grand Old Man of Dutch Liberalism." By forming an alliance with its former arch-enemies, the Catholics, the A.R.P. became the governing party. Schaepman, leader of Catholic unity, and Kuyper, leader of the conservative Calvinists, joined forces in their opposition to public education. Thus was welded a coalition that dominated Dutch politics up to the time of German occupation. They were able to effect legislation for government support of private religious schools but were unable to stop the march of social legislation that the liberal movement had initiated. This was partly due to Catholic sympathy for social and labor legislation. Catholics comprised the major element of the Dutch working class. Moreover, the Papacy had issued many calls for social justice and Christian democracy in the Western World.

The post-war period saw little alteration of these party alignments. The Occupation had altered many aspects of Dutch life, but not religious loyalties. There was, however, a change in emphasis. A kind of consensus had developed and religious and ideological dedications took second place to a renewed feeling of concern for the individual and national welfare. The Catholic People's Party (Katholieke Volkspartij – KVP) emerged in 1945 the nation's strongest single party. It is less concerned with religious particularism than with social progress. The wartime resistance experience of many Catholics led them into cooperation with other religious groups and away from an earlier pattern of Catholic exclusiveness. In the glow of brotherhood immediately following the Liberation, many questioned the need for a Catholic Party but no secular party system evolved and a Catholic Party somewhat on the model of Europe's Christian Democrats came into being. Today it seeks to be a more open party with less reliance upon the Church and with broader cooperation with other socially progressive groups.[1]

[1] Bishops no longer interfere with the Party or politics. The last such intervention

Since many of its aims for labor, justice and social welfare have been accomplished, the Catholic Party now concentrates its purely religious goals on obtaining better educational opportunities for Catholics and for greater economic development of the somewhat industrially trailing, largely Catholic provinces of Utrecht, North Brabant and Limburg. It favors increased public housing, full employment, more equitable wealth distribution and larger worker participation in industrial management. Its foreign policy is for European unification, reciprocal East-West disarmament, and economic aid to underdeveloped nations.

Much reduced from their earlier role, the Protestant Conservative Parties still remain. The ARP's rigid adherence to fundamentalist principles has lost its support in these modern times. Its faithful are largely members of the orthodox Calvinist Church (Gereformeerd) who assert that the State should remain subordinate to God. It insists upon state support for religious education. Though it calls for greater freedom of initiative for those in economic enterprise, it does not oppose social legislation already in existence. In fact its program calls for increased labor voice in industrial decisions and for increased European cooperation and foreign aid. A more liberal group, supported largely by Dutch Reformed Calvinists (Nederlands Hervormde Kerk) splintered away from the ARP in 1908 and has been gaining membership ever since. This is the more moderate Christian Historical Union (Christelijk Historische Unie – CHU) which has a program similar to the ARP. It also exhibits a similar dichotomy of aims. While calling for greater tax relief for the wealthy, it asks for increased social benefits for the citizen. Urging greater freedom of initiative at home, it speaks for more European cooperation and aid to underdeveloped areas. Each of these confessional parties has about seven to nine percent of the electorate's support. In the past the Catholic and Calvinist confessional parties, which represent moderate to conservative positions, have succeeded in capturing about half the nation's votes and half the seats in Parliament. In the last election their parliamentary representation was cut to one-third with their traditional opponent, the Labor Party gaining many of their seats and government controls.

Holland's major party is the Party of Labor (Partij van de Arbeid – PvdA). Like the KVP the PvdA was born in the 1945 post-war reor-

was in 1954 when a pastoral letter warned Catholics against joining Socialist Labor Unions. The Dutch Catholic hierarchy is notoriously liberal, challenging Rome on such points as lay participation in the Church, clerical celibacy and birth control.

ganization. Its forerunner was the Socialist Democratic Party (SDAP) which was founded by Domela Nieuwenhuis. Later Nieuwenhuis, impatient at the slow pace of social reform, moved away from faith in parliamentary reform towards anarchistic action. He had few followers however. The Dutch Socialist Democratic Party has always been reformist rather than revolutionary. During the late 19th and early 20th centuries it was ably guided as an opposition party by P. J. Troelstra. Today most of the socialist aims have been achieved. The Party of Labor is a fusion of Socialists, Social Democrats, and Social Christian elements. While secular in its outlook, it does not oppose the right of expressions of churchly viewpoints. It favors restrictions upon wanton exercise of property rights, a more equitable distribution of national wealth, and the defense of labor interests. Like all other Dutch parties it supports European cooperation and measures in favor of foreign aid. One of the best organized of Dutch parties, it participated in the post-War Government up to 1958 when it withdrew over a wages and income distribution dispute. It returned for a year (1965–1966) under the Cals Administration but it has been a party of opposition up to its 1973 victory.

Beginning in the late sixties, there has been a trend for new parties to arise at the expense of traditional party strength. The new parties are non-confessional and have their roots in the humanist and socialist traditions of the past. The humanist tradition is presently represented by the People's Party for Freedom and Democracy (Volkspartij voor Vrijheid en Democratie – VVD). Its membership is atypical of other parties weighed as it is in favor of wealthy recruits from the Dutch upper middle, professional, semi-professional classes. The Democraten '66 (D66), a somewhat social liberal party composed of younger, educated people also gained influence. Though a party in the present government coalition, it appears to be waning. A similar pattern has occurred with the Social Democrats (Democraten Socialisten '70 – DS70), a splinter group from the Labor Party which gained parliamentary seats in 1970 but lost ground in the 1972 elections. Another splinter group, left of labor but not communistic, is the Pacifist Socialist Party (Pacifistisch Socialistische Partij – PSP). A more significant progressive left wing party was the Party of Political Radicalism (Partij Politiek Radicalen – PPR). This party united with the Labor Party and D66 to form a progressive coalition in the 1972 elections. They are presently represented in the government. In addition to these, some even smaller parties have sprung up gaining one to four seats in parliament. Among them are two protestant parties (Staatkundig Gereformeerde Partij and the Politiek Gereformeerd Verbond): a splinter Catholic party (Nationaal Katholieke Partij) and a Peasants' Party (Boerenpartij) whose policy is based on national farmers and middle class interest.

Due to the Dutchman's insistence upon party ideological purity there has always been a superabundance of parties – a coming and going of splinter groups. In 1933 there were 54 party positions on the ballot. There were twenty-three on the 1972 ballot. There is even a Communist Party. It was quite acceptable in Holland immediately following the Liberation, garnering 11 percent of the popular vote. It rapidly lost support following the heavy-handed Russian Cold War tactics – especially events in Hungary and Czechoslovakia. It still has several seats in Parliament with its support almost exclusively from the working classes in Amsterdam and Rotterdam, and Groningen Province. Before the German invasion there was also a Dutch National Socialist Party under Mussert. In these manifestations the Dutch followed the pattern of their neighbors. Like the French, the Dutch have even developed a negative poujadiste-type party – the Boerenpartij.

The Dutchman's identification with either the Catholic, Calvinist, Socialist or Humanist tradition goes beyond the merely political. It is deeply rooted in a value system that is reinforced by church, union, recreational, informational and social associations that create a life style oriented towards one of these traditions resulting in the Dutch social phenomenon that Dutch sociologists call pillarization (verzuiling). There is presently a weakening of these pillars or traditions that had made Dutch society so stable in the past. The indecisive aspects of modern Dutch politics reflect this trend.

For 25 years following the Liberation the coalition of confessional and humanist parties dominated Dutch politics, with the labor and socialist parties forming the opposition. By the late '60's changes in the society were being reproduced in party representation in the government. Economic problems, primarily an accelerating inflation, arose to intrude itself as an issue in politics. Matters came to a head in the summer of 1972 when the Biesheuvel government seemed unable to effectively hold together its cabinet coalition. An election was called for November 1972.

The November election revolved around the issues of social permissiveness and the question of implementing government wage and price restraints to halt inflation. The confessional parties made their appeals to what they depicted as increasing social laxity – easy divorce laws, public pornography, the irresponsible actions of groups like the Kabouters and Provo's. The labor, socialist and DS 70 parties emphasized the issue of economic controls. Apparently this was foremost in the voters' minds. There was a noticeable shift of voter preference for the labor and progressive parties. There was decline in parliamentary seats for the Christian Democrats. The VVD, now more aligned with the left, gained handsomely, but the formerly promising D66 and DS 70 slipped considerably. The overall results were not conclusive.

The significant change may be less in the party alignment than in the type of men coming to office and the kinds of programs they are proposing. The new government has young and fresh blood. Many are young, only one cabinet member being in his 60's. Significant posts of Finance, Economic Affairs, Social Development and Education are headed by individuals still in their 30's.[2]

The Premier Joop den Uyl announced his coalition's main objective being "to reduce inequality in income, ownership, power and knowledge."[3] Such themes quite clearly are trends of the times. The specific governmental proposals for equalization of blue- and white-collar incomes, reduction of compensation of high salary workers, of tightening of price controls and profit margins have to mesh with the requirements of a growing economy, a control of inflation, and the continuation of government social services now so integral a part of the structure of modern Netherlands. All of this will be no easy task, even in as disciplined and responsible a society as exists in that country. Specific proposals for the creation of new advisory boards to help the government democratize its decision-making process are not necessarily going to improve "bread and butter" or life style conditions in the nation. Proposals for small reductions in defense appropriations to free sums for greater social services seem rather weak. Change may end up being more apparent than real.

Despite growing party divisions, Dutch government policies remain relatively consistent. This is because there exists a high degree of con-

[2] W. F. Duisenberg, 38 (PvdA), R. Lubbers, 33 (KVP), J. P. Pronk, 33 (PvdA), and Joseph A. van Kemenade, 36 (PvdA). Six other cabinet members are still in their 40's, and five are in their 50's.

[3] *New York Times*, May 31, 1973.

sensus on the priorities and directions of national programs and policies. An altered Dutch society came out of the humiliating and tragic experiences that all inescapably shared during the oppressive Occupation years. A new sense of social solidarity and dedication to humanistic purpose became a common conviction. The nation vowed that such experiences, or any form of human misery and degradation would no longer be tolerated. All agreed that fulfilling the needs of all the people was the first task of the State. Patriotic, religious, labor, professionals – even business groups – all echoed this theme. It was an altruism that the Dutch somehow managed to implement in a most pragmatic way.

Such common convictions simplified the path of post-war government policies. One avowed intention was to remain free from any future foreign domination. Consequently, the Netherlands abandoned its centuries-old tradition of neutrality and aligned itself with other democracies in defense of its freedom. It is a responsible member of NATO. A second policy was to support any program of international economic cooperation. This was an effort to help avoid a repetition of the economic hardships she had suffered during the Depression and War years. The Netherlands was the first to sponsor Benelux and other efforts towards a Common Market. Her present policy is to extend the scope of the Common Market and all other areas of free trade. A third consistent policy aim is the implementation of the government as the active promotor of economic and social welfare of its citizenry. Much of this, of course, implies socialism. The approach of the Dutch is fully practical and if capitalism serves, it is used. They have no overriding "hang-ups." They just believe that everyone ought to have adequate incomes, housing, health, education and environmental protection.

What was accomplished by this matter-of-fact, common sense approach to the nation's needs was the creation of a virtually new post-war society. Revival of the economy destroyed by the war, reconstruction of the nation's industrial infra-structure, equitable sharing of the nation's wealth, maximum health standards, decent housing for all, creation of a living environment – all these seemed so necessary and logical that there could be little controversy concerning them. Like the necessity to maintain the dykes and land reclamation programs, they have become an automatic aspect of government policy. The post-war consensus gave birth to the "Societal State."[4] Controversies that are

[4] The Dutch sociologist, P. E. Kraemer employs the felicitous phrase, "the Societal

growing more and more are over the issues of how most intelligently to implement the "Societal State." These decisions have been the concerns of professionals and technicians. Even the present Dutch society is directed largely by a core of men grounded in "expertise." With the democratization of interests and higher levels of public involvement in affairs, opposition to such powerful positions is rising. Still there is little of the "politics" we indulge in in our routine governmental affairs in America. The Dutch homogeneity of viewpoint and their application of technical and scientific management make much of it irrelevant.

This is amply revealed in the budgetary structure of the Netherlands' national government. A comparison of their budgetary priorities with our own is most enlightening. The Dutch have one unique item, a burden not imposed upon us. There is a fixed annual budget cost of approximately ten percent which goes automatically to control the sea and finance long-term reclamation schemes. Undoubtedly being accustomed to this inevitable expenditure and knowing the value of such public cooperative efforts made it easier for the Dutch to undertake further similar public commitments. Such "communistic" efforts destroyed neither their character nor their economy. The single item that looms largest in their annual budget is education. Along with expenses for recreation and cultural affairs, education takes up approximately 30 percent of the budget. One must understand that in the Netherlands school expenditures are not local but national burdens. The Government prorates allowances to local school districts and private and religiously based schools.

Excluding school expenditures, which in America are state or local budgetary items, the Dutch national expenditure on individual development and welfare is considerable. Public health and social services, along with public housing take a large share of government spending. Almost one-half of the budget goes for housing, public health, education, social insurance and allied social services. In our American order of priorities an equal amount goes for defense, debt interest, Veterans Administration and other costs allied to war. The Dutch allowance for defense, the most challenged item in the budget, is normally 15 percent.[5]

State" to explain this Dutch phenomenon in his provacative book *The Societal State*, Meppel, (J. A. Boom en Zoon) 1966.

[5] This is largely due to Netherlands' commitment to NATO. The Dutch have two mechanized divisions, one mobile division, with supporting units and cadre for expansion. Some units are already serving in Germany. It has eight airforce squadrons

Payment of interest on the national debt amounts to ten percent of the American annual budget. Dutch payment for interest on her national debt (a nation that was bankrupt in 1945) is less than one percent! The Dutch spend an additional six percent in financing public transportation and developmental programs in agriculture, trade and industry. Administration costs and the policing of justice each take about three percent more of the budget. This short survey reveals much about national attitudes and government management. The Dutch simply do not feel they can use their wealth unwisely or postpone indefinitely their financial obligations, even to themselves. The American's belief in his nation's inexhaustible wealth has made him exceedingly indifferent to how his government spends his income or what he gets in return for it.

The Dutch are apt to be more responsible towards the obligations of government, as seen in their acceptance of their tax burden. Annual tax levels approximate 34 percent of the nation's Gross National Income. Corporate taxes run 43 percent on profits up to about $ 12,000. They are mildly progressive beyond this point. An income tax for a typical middle class family of four would run about 30 percent of taxable income. Even with rising inflation and unemployment taxes are scheduled to rise. Local taxes tend not to be as onerous as in America. Real estate taxes approximate five percent of assessed values at the turn of the century. There are, of course, a host of other taxes – turnover, excise, dividend, wages, death, – that the flesh seems heir to. There is in the Netherlands, however, a relatively low level of tax evasion and fraud. Her tax structure is simpler, perhaps fairer, the government more scrupulous, the taxpayer more honest. There are no glaring tax bonuses like our Oil Depletion Allowance. Taxes are more truly progressive. It is unlikely that the Dutch middle class taxpayer will pay percentage-wise more than the millionaire, as seems to be often the fate for his American counterpart.

Not all government decisions are made at the top and much of the orderly functioning of society is eased by individuals in the lower, routine levels of government. The Netherlands has a tradition of devoted civil service comparable to England, Germany or Scandinavia. As in England the Dutch cabinet system creates a duality of political

and ten surface to air missile squadrons. The Dutch Navy consists of one launching cruiser, twelve destroyers, twenty corvettes and six submariners. Its contribution to NATO is considerable, however, there is growing opposition to defense expenditures.

and administrative functions. Governments come and go – the administration remains. There is no "spoils system" for all reaches of the Civil Service are outside the political arena. Dutch government employees are, relatively speaking, higher paid than their American counterparts and hold a higher status in society. Prestige in government service is such that individuals deliberately plan their university careers exclusively for it. Almost all executive posts require such training. The concept of the Crown helps in that civil servants feel they give their services and loyalties to a cause higher than a political party. Moreover they remain undisturbed in their tenure. Few personnel changes occur with a change in government – hardly more than among the Cabinet – and the newcomers are assured the same faithfulness and diligence given their predecessors.

There are approximately 400,000 national employees. Most are in the services – postal, harbor and customs, railroads, mint, national police, etc. Beyond these immediate service positions most government employees are in defense, education and the tax department. The Dutch postal service includes telephone and telegraph services which are government monopolies operating at a profit to the state. The Dutch would hardly for a second countenance the notion that so routine a function as the postal service would merit a cabinet post. About ten percent of the nation's work force is employed by the government, including over twenty percent of its white collar workers. In America, which has only half that level, the public would react with horror were we to employ that many. Most Americans feel that their 3,500,000 federal employees already have too little to do. Yet they overlook the hidden subsidy represented by the one out of ten jobs dependent upon our federal defense spending. Unfortunately our military-industrial complex method of income distribution does not work well for wider wealth sharing. High wages are paid technicians and high profits accrue to limited space-defense industries. Government monies poured into urban renewal or land reclamation, would open up more jobs for the unskilled and involve a wider industry coverage. The Dutch government looks candidly upon its Civil Service as a form of income redistribution and many menial government tasks serve as substitutes for direct welfare. The government in the past has successfully initiated projects for canals, roads, railroads, textile industries, among others, as much to overcome unemployment as to foster industrial development.

PROVINCES OF THE NETHERLANDS

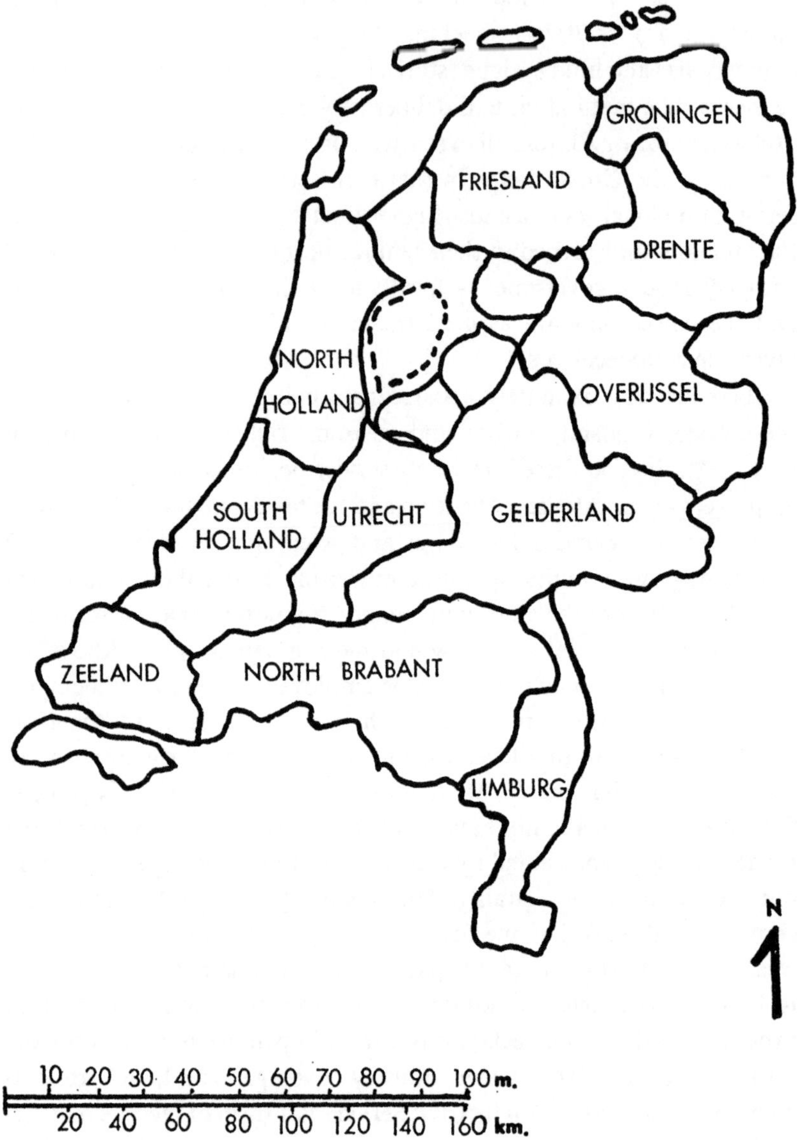

The Dutch arrangement for lesser government units is not federal but unitary. The national government plays the dominant role throughout. In a nation of the size and homogeneity of the Netherlands, such centralization is a logical inevitability. There are eleven Provinces that serve as intermediaries between the National and Municipal governments. These are South Holland, North Holland, Utrecht, Zeeland, Gelderland, Friesland, Groningen, Overijsel, Drente, North Brabant and Limburg. The new lands created from the IJsselmeer, presently under national administration, will ultimately form a twelfth province. Their functions are concerned primarily with traffic, public utilities, cultural development and environment control. They more closely resemble our counties than our states. Provinces are governed by an elected Assembly (Provinciale Staten) of from 39 to 83 members depending upon provincial populations. The real power resides in the Queen's Commissioner who receives his appointment and instructions from the Crown and is responsible for justice, good order and provincial compliance with national directives. Provincial funds come largely from the national exchequer. Only one-sixth of provincial expenditures is covered by provincial taxes. Still the Provinces receive only about five percent the amount that the National Government gives to the Municipalities. This is barely as much as that given to the Waterschappen, the administrative units responsible for water protection and control.

Of much greater importance is the role that Muncipalities play in government. The Municipalities (Gemeenten), particularly the larger ones, are still powerful. This is an outgrowth of the medieval traditions when cities were centers of democratic, burgher power, standing out against the encroachments of surrounding, and often foreign, feudal lords. The rights of the cities to self-rule have been jealously guarded but there has been a trend towards centralization of power following the Napoleonic Occupation and the 1848 Constitution. Under that Constitution Municipalities were given a uniform system of government. There are approximately 900 Municipalities running from 300 to 900,000 in population. The final political responsibility for government rests with the elected Municipal Council, of 7 to 45 members, depending upon population. A smaller body, the College of Burgomaster and Aldermen, serves as an executive body for day to day affairs. The real power often rests with the Burgomaster, again a Crown appointee, and a career administrator. In a sense the Municipalities are run under city management, with a nationally selected Director speaking for the national interests. The Burgomaster has a suspensive vote over any ordinance passed by the City Council. In a conflict the Municipal Council can prevail. They can even dismiss the Burgomaster, though his normal period of tenure is for six years. However, if he is capable, the Burgomaster can persuade the Council to his views. He is, in many cases, university-trained in Municipal administration and looks for a successful and progressive record to recommend him to a more challenging, prestigious and lucrative, large-city appointment. Councilmen are themselves capable community leaders and have, generally speaking, considerable experience in local affairs. They have usually shown themselves open to rational plans for municipal management for development when cogently presented to them. The usual result is an acceptable level of professionalism and efficiency in local government.

The Municipalities are responsible for the traditional functions of police and fire protection (police services may be provided for by the National Government, and the fire department may be volunteer in the smaller communities), public utilities, public health, and sanitation, housing and public morality. There are many national laws, such as alcoholic control, building codes, health regulations, etc., that hold the Municipalities to task. Moreover the lion's share, approximately 90 percent of Municipal funds come from the National Government. There are few problems of public order or maintenance of public services. Trash accumulations do not occur, and but for beginning evidence of neglect in the working class areas of the bigger cities they are rather clean. Cities are made attractive with mini-parks and plantings that are normally left undisturbed by the inhabitants. Ordinarily one is safe walking about the cities at any hour. Since welfare and public health are national concerns, the most serious municipal problem is public housing. There is a growing tendency due to increasing populations as well as the creation of regional authorities where mutual interests are involved, for smaller communities to consolidate or to enter into cooperative ventures in public utilities, water or sewage projects, or public transport. The two outstanding regional arrangements are the Rijnmond and the IJmond – port and industrial complexes for the Rotterdam and Amsterdam regions. There is also some consolidation of the Waterschappen or Water Control Boards.

Dutch cities have had histories of successful economic management and compassionate concern for their poor and unfortunate. Undoubtedly these traditions explain to some extent the more Democratic management of urban problems that the Dutch have achieved when we compare their approach with our own conspicuous urban failures. In all fairness we must appreciate the historic, sociological and other differences between the two cultures. The Dutch never experienced the strain of the "melting pot" task of assimilating a massive alien immigration. Nor have they been afflicted with the dilemma of race that besets our society – particularly our overburdened cities. Theirs has been a homogeneous race and culture. Race and ethnic animosities are surfacing primarily in the larger cities of the Netherlands. Some 100,000 Indonesians were accepted into the country when they opted for loyalty to the Dutch during Indonesia's rebellion against colonial rule (1945–49). The Surinamese, Arubans, and Amboinese – emigrants from the Dutch colonies – constitute additional minority elements in a previously all white society. The influx of foreign workers – Spaniards, Yugoslavs, Algerians, Turks, etc., – further test Dutch levels of tolerance towards outsiders. It cannot be said that such a test passed with flying colors. Still the level of intolerance has not quite reached the intensity exhibited by British opposition to their colored Commonwealth immigrants. Nor is it conceivable that racial conditions in the Netherlands will ever quite parallel those we find at home. There are clear social disabilities levelled against alien ethnic groups, but on the other hand, wage differentials, housing discrimination, or outright humiliating hostility, in general, in no way matches what is typically found on the American scene. For example, the Dutch would hardly conceive of laws banning inter-racial marriage. In fact, such intermarriage is rather frequent and is socially acceptable.

Racial conflict in our cities and the dangerous contrasts of affluence, cheek to jowl with poverty, are of our own making. Yet the tragedy of animosities of race and class can be somewhat accepted as understandable human failings. Some of our other notorious urban shortcomings cannot. Housing shortages, transportation failures, costly and inefficient public services, crime and corruption – in short, urban decay – still remain. The Dutch tradition of honesty and efficiency in municipal government still contrasts with our own history of municipal mismanagement, blackmail, graft and jobbery. Police corruption, so commonplace with us, as the recent New York City Police Department investigation underlines, is virtually unknown to them. American cities did not become corrupt and bankrupt overnight, as Lincoln Steffen's *Shame of the Cities* well pointed out over 65 years ago. It is our way of doing things. Dutch opinion is inclined to view our urban plight as justifiable. The Calvinist ethic has little patience for our variety of civic irresponsibilities.

In general, one can speak well of the Dutch bureaucracy. Its Civil Service, including Provincial and Municipal levels, offers rewards and status that enhance the calling. Undoubtedly those holding higher and more interesting offices are truer to the Civil Servant model. Lesser functionaries may be less careful, perhaps more indolent. Assuredly much of civil service is routine, boring, bureaucratic. Yet there is a business-like air in many government Kantoren. To the Dutch no task that is honest can be demeaning. One encounters a certain pride in work, order and cleanliness. However with the growing affluence and sophistication of Dutch society there appears a seeming erosion of these virtues.

CHAPTER III

MANAGEMENT OF AN ECONOMY

We have previously characterized Holland's ecology as a dense population bearing upon limited space and resources. Her geographic limitations have been compounded by the recent events of a depression, a military occupation, the dismantling of much of her industrial plant, the loss of established trade channels, and the end of her colonial empire. Yet by "force of methods" Holland has attained an economic recovery that well deserves the term miraculous. Her national income rose from $ 4 billion in 1949 to $ 65 billion in 1974.[1] Undoubtedly the root cause of Holland's success has been the general rise in world prosperity. A nation whose economy is dependent for over half its income on world trade could never gain such achievements alone. Yet the formulation and coordination of deliberate economic policy aimed at full employment, maximum capital development, and high-level trade balance, could not but help contribute towards the rising national incomes that the nation experienced.

The formulation of economic policy in the Netherlands is a coincidence of a political process, which determines policy objectives and means to achieve them, with a technical process of information collection and analysis that serves the political planners. In the process, it is doubtful that any other Western country has gone as far as the Netherlands in attempting to forecast economic levels and in evaluating the economic consequences of government policy decisions.[2] The

[1] Even discounting the inflationary rise during this time, this is a considerable achievement. Much of the recent increase is due to the re-evaluation of the Guilder in response to the official dollar devaluations of 1972 and 1973. The first saw the exchange rate of the guilder drop from 3.6 to 3.2 to the dollar. The second saw a further drop to 2.9 to the dollar. It has gone as low as 2.4 to the dollar.

[2] James Albert, *The Coordination of Economic Policy in the Netherlands*, (Duke University Press) Durham, N. C., 1966, p. 111.

purely political institutions involved are the Crown, Parliament, Cabinet and the political parties. The political bodies hold ultimate responsibility for economic planning decisions. The technical institutions involved are the Central Economic Planning Bureau, the Central Bank and the Central Bureau of Statistics.

The major trend in economic policy planning in the Netherlands has been and continues towards greater delegation of responsibility for planning by professionals. There are many reasons for this. With the exception of the Labor and new liberal parties, the Dutch political scene was organized on confessional party bases. The issues important to the confessional parties, were and remain, rather irrelevant to social and economic planning. The majority Catholic Party had some interest in working for the goals of social and economic justice outlined in the Papal Encyclical Quadragesimo Anno 1913. The driving force for economic planning came traditionally from the Labor Party which hoped for a larger role in government and society. This was particularly true at World War II's end. The Labor Party found the ready cooperation of other groups, particularly the Catholic Party, and together they launched those measures that set the Netherlands upon the road of near total social and economic planning. There had always been heavy historic elements of bureaucratic paternalism in the Dutch government. The Crown on its own initiative played a significant role in early economic planning. Later the expansion of popular democracy led to disturbing conflicts within society and between the emerging political parties. The Dutch were eager to avoid conflict and turned to the impersonal bureaucratic Crown professionals in working out their problems. This tradition continued to modern times. Also important in making for further professionalization of economic planning is the nature of the Dutch economy. It is highly dependent upon international trade, the details of which are so technical that they remain outside the political process, except for the designation of desired goals. Moreover, by creating technical bodies, political leaders are able to avoid political pressures and political embarrassments. All decisions could be referred to appropriate and politically sterile planning bodies.

Economic problems were plentiful enough in the immediate postwar years. Fortunately there was an amplitude of available "experts." Mr. P. Lieftinck, former Professor of Finance and Economy at the Economic School of Rotterdam, was appointed Minister of Finance from 1945 to 1952. Two other professors of economics, J. M. van den

Brink and J. Zijlstra, served as Ministers of Economic Affairs from 1948 through the 1950's. Dr. M. W. Holtrop, another economist, was Director of the Central Bank. The internationally known economist and statistician, Professor J. Tinbergen, was appointed Director of the newly created Central Planning Bureau.

Economic planning can be divided into four major periods, during which the government had four different planning goals and employed different approaches. The first period ran from 1945 to 1950 and saw the government as a virtual dictatorship over the unreconstructed economy. Even here, however, the government relied heavily upon its advisory bodies, particularly the Foundation of Labor. This was a period of willing cooperation of labor and management with government aims and measures. All were bent upon the mutual goal of reconstruction and understandably tolerated the necessary restraints and inconveniences related to economic recovery.

The primary governmental instrument to effect government policy was the Board of Government Mediators, founded by the 1945 Extraordinary Employment Relations Decree. It consisted of six to eight members appointed by the Minister of Social Affairs, usually because of their experiences as wage negotiators. They were not civil servants. As its name implies, the Board concerned itself primarily with labor relations and wage policies. Labor, particularly wages, is the most important variable in Netherlands industrial production. Price and quality determine the marketability of products abroad and this is very important to a nation, one-half of whose income is derived from foreign trade. Increased trade meant capital for industrial investment and full employment. Disciplined, productive and reasonable labor was basic to achieving the low costs for a favorable export trade so vital for Holland's economic recovery. Fortunately labor understood that living standards and luxuries could wait. The Board of Mediators had near dictatorial powers over wages, conditions of labor, guidelines for collective bargaining – and exercised them completely and effectively. It did so with a minimum of governmental interference and with little political party conflict, and largely through the cooperation of labor.

The Board of Mediators, even under the emergency situation of post-war recovery, preferred cooperation rather than unilateral decrees and it worked harmoniously with the Foundation of Labor. The Foundation was a unique organization of Labor and employers that developed out of the Occupation when significant labor and business leaders cooperated in the Resistance Movement. Its founder was Dr. D. U. Stikker, former director of the biggest Dutch brewery, later ambassador to the United Kingdom and recently retired Secretary General of NATO. The Occupation had revived Dutch nationalism, furthered labor solidarity and heightened the hope of fraternal cooperation of all elements of the economy following the Liberation. There was a touch of the utopian in its early aims.[3] Though union registration has always been quite high in the Netherlands, labor has rarely been militant. It has been only with the seventies that Dutch unions became more seriously directed towards wage and hour issues approximating the American pattern. The Foundation of Labor was more generalized representing all labor groups and from 85 to 90 percent of all employers. Somewhat undermining labor solidarity was its division into religious categories following the pattern in almost every aspect of Dutch life. The three most significant organizations are the Netherlands Federation of Trade Unions (secular), the Netherlands Federation of Catholic Trade Unions, and the National Federation of Trade Unions (Protestant). In 1969 their membership was 555,000, 489,000 and 240,000 respectively.[4] Workers and employers in agriculture and smaller businesses are less apt to be union members. All major concerns in the Netherlands belonged. Immediately after the war the government adopted the practice of consulting the Foundation on all labor issues. Until the passage of the Industrial Organization Act of 1950, the Foundation was the principal government advisor on labor and wage decisions and virtually controlled labor conditions in the Netherlands.

The industrial Organization Act created the Social and Economic Council (Sociaal Economische Raad) and the government shifted its policy from tight controls and reconstruction towards effecting a more equitable distribution of now tangible economic recovery. This brought on the second period of economic planning dominated by concern for labor interests. It was an effort to give some reward to labor after the belt-tightening it accepted during the time of economic reconstruction. The Economic and Social Council was designed to steer the economy on this new course. The Council was composed of 45 members, one-third from the ranks of labor, one-third employers, the last third

[3] Albert, *op. cit.*, pp. 43–44.

[4] *Netherlands, Work and Prosperity*, Pathuis, H., and van der Spek, J. Lochem, (De Tijdstroom) 1971, p. 21.

government Crown appointees. The government representatives were largely independent experts – scholars of law or economy. From the beginning these government appointees tended to be the "swing vote" in deadlocked issues between labor and management. They also tended to lift decisions out of the political arena. The Social and Economic Council was very powerful for its decisions had the force of law. Its professionalism made it a respectable body and its decisions were acceptable to most parties. This tended to perpetuate the institution of "buck passing" but it also served as a means by which government policy decisions could be explained, understood and finally accepted by those trade unions and employers organizations whose support was so vital for the implementation of government policy.

The Council's method to enlarge labor's share in the distribution of the now considerable economic growth was a wage policy to provide "equal pay for equal work." This was an attempt to standardize wages in work categories regardless of the earnings of particular industries. It was a policy of wage scales and had the hardy endorsement of labor. Its implementation meant the granting of government subsidies to those marginal industries normally categorized by low wages. The earlier policies of full employment, support of price levels, balance of payments and capital investment continued unchanged. The aim of the Industrial Organization Act to establish a comprehensive statutory framework for all branches and levels of industry under the control of the Social and Economic Council similar to what was attempted under our own National Recovery Act experiment, was not achieved. Since most industries kept the older organization structures and negotiated with separate trade unions – confessional and non-sectarian – the changes were largely paper ones. Industries failed to establish the commercial and industrial boards the government had planned for. Perhaps the government scheme required an over-organization that even the tidy Dutch could not accept. Though the Social and Economic Council had not yet become the apex of this proposed economic organizational pyramid, it did replace the Board of Government Mediators and the Foundation of Labor as the main body the government consulted for its decisions.

A third shift in economic policy occurred in 1958. By then economic recovery was complete and the danger now lay in the disturbing inflationary trends developing out of Holland's prosperity. Inflation had been growing since 1953 but had now reached alarming proportions.

Unemployment had dropped to 1.5 percent of the total available labor pool placing extreme pressures for government approval of labor's increased wage demands. The shortage of labor was forcing industry to resort to "black wages" and other devices that undercut the government's efforts to halt wage inflation. Labor leaders did concur with the 1957 government proposal for wage cuts, but not with the wholeheartedness with which they supported the 1951 decisions to cut back wages to combat the Korean War inflation. Dutch foreign trade was being affected by increased prices. The balance of payments had deteriorated from a surplus of $ 215 million in 1955 to a deficit of $ 200 million in 1956.[5] National income in 1956–57 rose seven percent, consumption ten percent, public investment 12.5 percent and inventories by 20 percent. All indications pointed to difficulties and called for drastic anti-inflationary measures.

The government requested the Social and Economic Council to offer solutions. It took the appropriate but politically unpalatable course of cutbacks in expenditures and wages with increases in taxes and prices. It recommended a $ 200 million reduction of government expenditures (one-half in government capital investment), a cut in local expenditures, withdrawal of investment allowances to corporations, raises in income and corporate taxes, hikes in rents and in the cost of social insurance through payroll deductions. The result was a cost of living rise of 12 percent but a favorable balance of payments of $ 112 million.

The difficult problem of wage policy remained. If a balance of payments was achieved in international trade, if high levels of capital investment continued, the resultant prosperity, desired but inflationary, would cause full employment, increased consumption and heightened pressures for increased wages. The government attempted to cut the Gordian knot by changing its wage policy from one of equalized wage scales to that of differential wage rates. It wanted industry to seek out and permit a level of wages upon the actual productivity of labor. This policy was vigorously resisted by the Labor Party which retired from its participation in the government over the issue. The Labor Party-Catholic Party coalition was replaced by a new coalition of the Catholic Party, two Protestant parties (the Anti-Revolutionary Party and the

[5] *f* 750 surplus in 1955 to *f* 710 deficit in 1956. The Dutch Guilder had a current exchange rate of approximately $. 28 at that time.

Christian-Historical Union), a coalition that, with some interruption, persisted up to 1972. Cabinet difficulties resulted in an election that year – difficulties that have as yet to be overcome in Parliament.

This controversial wage policy was in a sense an attempt to get back to normal – a free, openly competitive, domestic economy. The government employed the existing machinery to regulate its proposed changes. Collective wage bargaining began at local levels of industry, at each separate corporation. The Board of Government Mediators was retained to administer wage policy though under the control of the Social and Economic Council. Since most corporations and labor organizations were hardly equipped to realistically compute productivity levels, the government's Central Bureau of Statistics was enlisted to work out the technical aspects of the new policy. There was much evidence that labor and management usually decided upon their agreed wage hikes and then produced statistics to justify them. The Minister of Economic Affairs held many "price discussions" with labor and industry to hold wages and prices in check. Under the general inflationary trend of these prosperous years government hopes were largely defeated. "Black wages," fringe benefits and other subterfuges were practiced. Still the government acted in the direction necessary to curb inflation despite the political repercussions that went so far as to see its fall.

By 1963 wages were clearly outstripping productivity increases. The Government had reduced the work week from 48 to 45 hours in 1962. Labor costs that year rose 6.5 percent alone. The government once again turned to the Social and Economic Council. It recommended that the government switch over to the Foundation of Labor to effect its wage policy. In effect, the Council advocated its own demise. The government's adoption of this suggestion initiated its fourth phase of economic policy. It was a continuation of the attempt to return to "normalcy" by other means. There was a greater use of professional economic services and controls. In its battle to combat a heated economy the government thrust the responsibilities for controls upon management and labor – the elements of the economy that had to live with them. The Social and Economic Council became an advisory body to the government with the Foundation of Labor playing a more important role.

The Council continued publishing its September and February Reports that served as guides in determining yearly shifts in government policies. The Foundation discussed details of government aims with labor and employer organizations, worked out compromises and attempted to gain support all around for the government position. If no agreement seemed possible through these efforts, then the government, through the Board of Mediators, imposed its own wage levels – usually a wage freeze. The only recourse labor or management had was through parliamentary pressures upon the cabinet or the government. Usually dissident parties bided their time and made their wishes felt in the next year's planning projections.

The initial results of this new government policy was that there was some hold upon inflationary wages and prices. However, per capita wages gradually exceeded productivity levels. Labor leaders were no longer as willing to exercise restraints as they had been in 1945, 1951, or 1957. Many large firms were prepared to pull out of the Foundation of Labor. The Government had no recourse but to cede to demands for increased wages, which led inevitably to price rises. Prosperity and employment were too high for anything but a flexible wage and price control policy. Only increasing prosperity and the Netherlands' favorable balance of trade made inflation endurable. In 1965 wages increased by 12.4 percent, outstripping productivity levels. The same year saw a slight imbalance in trade, but the government seemed not too concerned.

There was, however, a growing stagnation of real economic growth by 1966. This occurred not only in Holland but also in most of the industrialized nations of the world. However, a recovery of economic growth developed in 1967, gained momentum in 1968 and continued through 1969. Gains were high in the Federal Republic of Germany, France, Italy, Holland, and above all, in Japan. There was a lag in the economy of the United States and of the United Kingdom. The Netherlands shared in the general West European prosperity, albeit one fueled by inflation.

Still Dutch prospects for 1969–70 looked optimistic to the Central Planning Bureau which predicted a decline in unemployment, but not in wage increases. Its predicted wage hike of ten percent was to be offset, unfortunately, by only a projected four percent increase in productivity. It anticipated that higher import prices, higher wages, and sluggish growth of productivity would have adverse effects upon the nation's balance of payments. Predicting a 7.5. percent cost of living rise, the Planning Bureau advised against the easing of credit despite the year's tight money supply. There had been government cutbacks in capital spending partially to cool the economy. There was a slightly higher rate of growth of unemployment than was anticipated. While not accurate in all its predictions, 1969 proved to be a comfortable one for the Dutch.

The Central Planning Bureau forecasts for 1970 were most interesting for this was the year of great economic turbulence, particularly in the United States, the pace-maker of the world's economic trends. The Planning Bureau's September 1969 report felt that most industries throughout the world had reached the limit of their productive capacities, particularly since demand growth had been so great since the 1967 recovery. The attempts of numerous countries to curb inflation and wage-price spirals would, in its opinion, contribute to lagging investments. While predicting this would reduce somewhat the pace of inflation it also saw how it would reduce world trade. Reacting to these events the Central Planning Bureau predicted a "cooling off" of the rather overheated economy the Netherlands had been experiencing. It forecast that the cost of living would rise only half as much as the year earlier and that wages and productivity increases would be in balance. If continued, this would reverse a trend that was heightening inflation and creating a dangerous international trade situation for the Netherlands. Perhaps encouraged by these signs the government finally advocated the firm policy of economic and fiscal controls suggested in the 1970 Central Planning Bureau Report. This included cutbacks in government expenditures, an increase in the income tax and, over the protest of labor, only 5 percent wage increases. Highly unpleasant to the electorate, these measures represented planning and control, not mere "jawboning" of the American model.

It is important to note that the Central Planning Bureau in its appraisal of the Dutch economy sets it firmly in an evaluation of world economic conditions, something our own economic forecasts are apt to overlook. Perhaps, in this connection, the most illuminating part of the Bureau's Report was its analysis of the American economic situation in 1969 and its prospects for 1970. It predicted that inflation in America, which was serious in 1969, would be rampant in 1970. It reported that the American Gross National Product which nominally rose by six percent the first half of 1969 was, discounting this inflation, a bare 2.5 percent – one of the lowest growths of any industrial nation. It further predicted savings declines as private debt consumption increased ... "in absolute terms the increase in consumption by households greatly exceeded the increase in disposable income ..."[6] It concluded by predicting a continued rise in inflation levels throughout American industry.

Nor was it wide of the mark. Real inflation in America in 1970 proved to be slightly less than one percent below the Report's prediction. American consumer debt was at an all time high in 1970. Industrial price rises were high but were matched or exceeded by costs, particularly wages. In fact much of the year's investment was for capital machinery to replace highly paid and often productive labor. 1970 was not a good year for American industry.

The Central Planning Bureau Report was remarkable in its thoroughness and the accuracy of its diagnosis. Though couched in polite terms, the Report had said, in effect, that America, the world's wealthiest nation is in the process of spending itself into bankruptcy. Consumer debt has risen and savings have been consumed. The market has been over-sold. This explains the present conjunction of tight money, high interest rates, growing unemployment and inflationary prices in our economy. It also explains the lower profit levels on increased volumes of business. The Spring 1970 decline in the American stock market was a reflection of these trends.

The Dutch economic forecasters, and those of Europe, generally, were taken by surprise over Nixon's 1971 wage and price "freeze" and his foreign import surtax. Still, based on our continued indifference to management of our economic difficulties in the past, it is understandable that Dutch economists never conceived of the American government as being capable of taking serious steps. In one sense their instincts were correct, for it is doubtful if the American government will adopt economic controls except as a stop gap measure.

While presently inconvenient the events of 1970–1971 probably will not be totally disastrous to the American economy. Seemingly, our superfluity of wealth permits us to be heedless of our economic difficulties. Normally when consumer demand revives after a period of underproduction, our industries need only to return with customary fury to their exploitation of our abundant natural resources. Whether our resources make it wise to do so, our economy is such that we literally eat our environment. Mineral resources of our former unspoiled coun-

[6] *Macro-Economic Report 1970*, p. 13.

tryside are now strung out along our highways, in junkyards and used car lots. Our former forests now blow listlessly as paper plates across the wastes of Nevada or collect as unread newspapers in the empty lots of Newark. The Netherlands exploits the resources of other lands. Her industry processes the resources of other people, selling them back at a profit. This explains why world trade is so important to her and of lesser importance to us. Her poverty has forced her to be diligent, thrifty and industrious. Our wealth permits us, when it doesn't actually force us, to build an economy of waste. In the process we are content to allow the ups and downs of our economy "to take their natural course." The Dutch instead attempt to control this as other matters and thus keep their economy on an even keel. Yet, as our present trade imbalance, dollar devaluations, and accelerating inflation indicate, our non-management of our economy cannot go on indefinitely.

A discussion of the professional planning service now becomes necessary. There are two technical institutions responsible for economic planning in the Netherlands. They are the Central Planning Bureau and the Central Bank.

The Central Planning Bureau (Centraal Planbureau) was established by the Industrial Organization Act of 1945. Prior to World War II the government of the Netherlands had at best a haphazard policy concerning the compilation of information concerning the nation's present and future economic prospects. The Depression Years pointed up the benefits of economic predictability and control. The Occupation postponed such intentions. The economic difficulties of reconstruction rekindled government interest in a central organization that could help it in its effort to organize investment and production, lift levels of productivity and income, and achieve fuller employment. The Central Planning Bureau was designed to render technical assistance to the government in its coordination of economic policy. It had no executive function; nor has it today. It is primarily a staff arm of the Social and Economic Council. It is protected from government interference in its professional function by the Central Planning Committee – a semi-official body composed of 11 government, 16 worker and employer, and 3 professional appointees.

The growth in size and influence of the Central Planning Bureau has followed the pattern of almost all the other planning and control agencies of the Netherlands. The Bureau began with a 75 man staff, half

THE POSITION OF THE CENTRAL PLANNING BUREAU IN THE GOVERNMENT OF THE NETHERLANDS

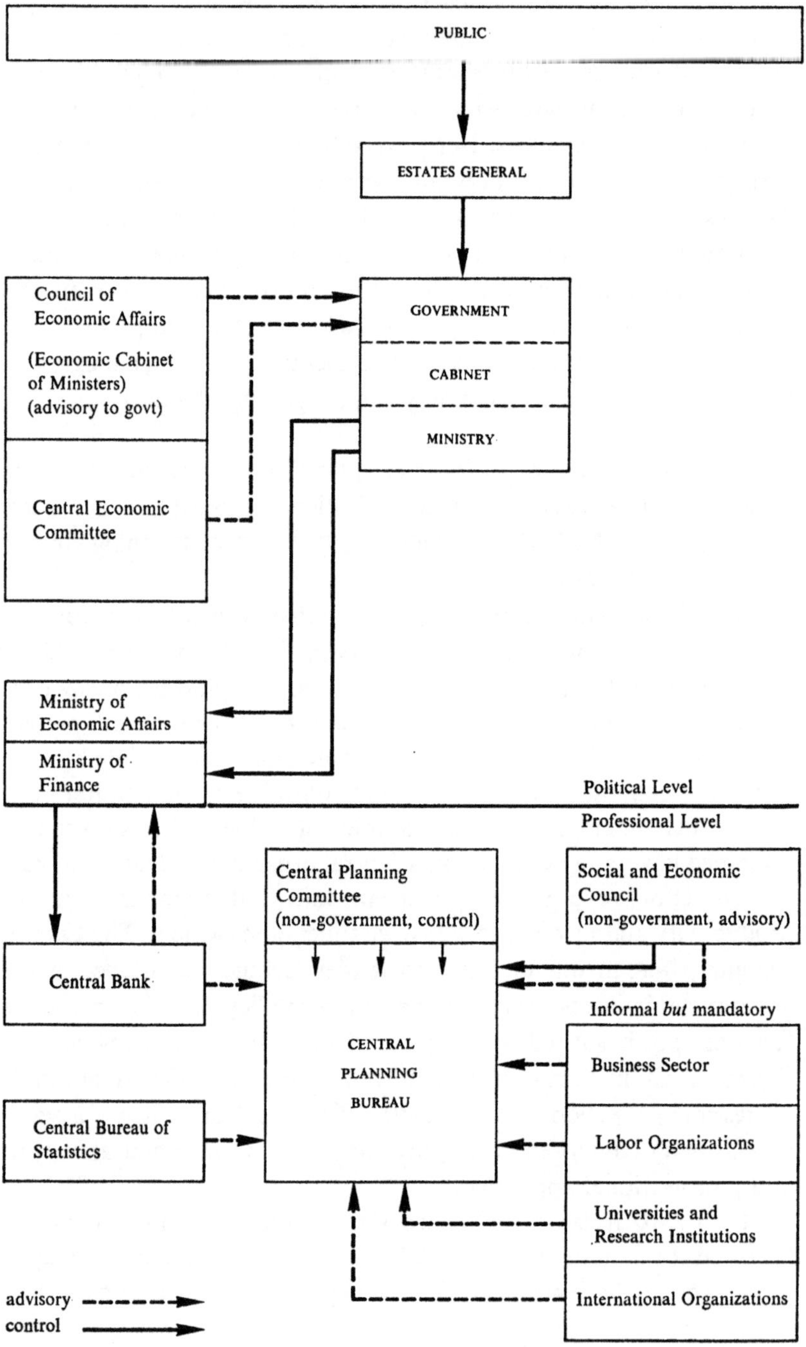

of which were university trained. It presently has over 115 university trained employees alone. These include a high proportion of mathematicians, statisticians, economists, engineers and a handful of professionals with legal training. Its latest reorganization has divided the Bureau into four technical divisions plus a household division with administrative personnel, library, documentation, drafting, computer operation and clerical functions.

The first technical division of the Central Planning Bureau is highly specialized, serving as "the keeper of the model." Here mathematicians, statisticians and economists work on the construction and improvement of various econometric models and continually attempt to improve their predictive accuracy. Mathematical formulas designed for making economic prediction, econometric models, offer a means of reducing complex economic relationships to some measure of manageable proportions. This is particularly necessary if one is to devise a scheme of economic planning on a national scale such as is attempted by the Central Planning Bureau. The earliest model was developed by Professor Tinbergen. It contained approximately 38 equations – 11 behavioral, 27 technical variables – of economic functions. The present model contains dozens of pages of technical and abstruse mathematical symbols devised to cover variables in the Dutch economy. Models are frightening and awesome to the layman but can prove to be surprisingly accurate when used by skilled economists. For example in an *Industry Memorandum* the government predicted that between July 1948 to January 1953 the demand for new positions for labor thrown out of work by the increase in labor productivity alone would be 120,000 or 23 percent. This prediction made five years ahead proved to be only a fraction under the real figure.[7]

[7] Westrate, *Economic Policy in Practice: The Netherlands 1950–1957*, (H. E. Stenfert Kroese B.V.) Leiden, 1959, pp. 23–24. Pioneer work on econometric models was done by L. Walras (1834–1910), French born professor of Mathematics at Lausanne University, who asserted all economic data could be reduced to mathematical terms. The econometric models of the Central Planning Bureau are based on those of Cobb-Douglas. The most frequent inadequacy of the models is that they cannot provide for all variables. Moreover subjective aspects of economics are not reducible to mathematical measurement. That venerable dean of economics, J. M. Keynes once wrote a criticism of Tinbergen's models. (J. M. Keynes, "Professor Tinbergen's Method," *Economic Journal* XLIX, September 1939) The most thorough discussion of the Central Planning Bureau's econometric models is to be found in Albert, James, *op. cit.* Chapters IV, V, VI. There is a constant need for experts to reassess the model and to check upon its accuracy coefficient.

The second technical division deals with short term planning. It has subordinate divisions concerned with the various effects changes in budgeting, production levels, interest rates, balance of payments, domestic consumption, wages, prices, rents, fiscal and monetary measures have upon the total economy. This division is responsible for preparing the annual *Central Economic Plan* that is the heart of economic planning and control of the Netherlands. A third division deals with sector planning – agriculture, industry, service, energy, transport. A fourth concerns itself with regional aspects in the economy – along with technical projects for regional improvement, equitable income distribution and social welfare programs. This last division works in close association with the National Physical Planning Service (Rijksplanologische Dienst) of which more will be said later.

The key to economic planning in the Netherlands is the *Central Economic Plan*. This is less a plan than a collection of figures relating to size of investment and production, levels of wages and prices, productivity and consumption, national income, balance of trade, and other data important to the coordination of economic, social and financial policies. Predictions are given on domestic and international trends. Its projections cover export opportunities, price range of imports and exports, labor supply and wage levels, expected government income and expenditures. It draws conclusions concerning them and outlines the results of government actions that might be taken. It often offers alternative suggestions for the government planners.

The sequence in preparing the plan is important to the final outcome of government policy. In May or June when the government is preparing its budget, the Central Planning Bureau works on a tentative economic report. The Bureau cooperates with the ministries in the formulation of their fiscal and budgetary intentions for the year. It makes formal and confidential suggestions to appropriate ministries and the government. There is mutual feedback and exchange of views. Discussions ensue in the Council of Economic Affairs around the revised budgetary positions. The process is such that the technical experts of the Central Planning Bureau wield considerable influence upon government policy decisions. The Council of Economic Affairs is responsible for preparing the budgetary position of the government for its presentation to Parliament.

The first published "Preliminary Plan" appears with the publication of the budget in early September. This Central Planning Bureau publication usually discusses quite frankly the economic consequences of the proposed budget. The rough draft of the "Preliminary Plan" is elaborated upon and brought up to date following parliamentary discussions and decisions on the budget. The result is a second draft which is laid before the Central Planning Committee, which reviews it primarily for its technical accuracy. Meanwhile the staff of the Central Planning Bureau has gone into the field to sound out important industrial executives and labor union leaders. It gets the "feel" of what these important segments of the economy are hoping to do. Final changes are made and it is submitted to the Council of Economic Affairs. Some revisions may come out of the Council of Economic Affairs but then it is finally published. This is usually in February of the plan year. This procedure illustrates clearly that the plan is based on government policy and not the other way around – basically a forecast of the effects or results of government policy. Experts of the Central Planning Bureau have great influence during planning stages of government policy. But they do not make the decisions. Executive power lies with the Cabinet, upon the advice of the Council of Economic Affairs and the ultimate approval of Parliament.

In addition to the annual Central Economic Plan the Central Planning Bureau prepares numerous special studies or reports requested by the government – Ministers of the Cabinet, Social and Economic Council, Council of Economic Affairs, etc. It has even prepared reports for the European Cooperation and Development Community on such diverse aspects of the economy as the social returns of large public investments, or the demand levels for future professional skills in the Netherlands. It makes continued studies and forecasts of the economies of the rest of the world. Since over 50 percent of the economy of the Netherlands is dependent upon international trade, this is almost a necessity. The Central Planning Bureau makes frequent business cycle estimates and forecasts their effects upon the world price indices. It investigates not only these aspects of its own economy but the changes in disposable income, order-to-sales ratios, inventories, profits and labor cost studies within the United States, Federal Republic of Germany, United Kingdom, France, Benelux, etc. The Netherlands is never permitted to ignore world events and world economies. Her welfare hinges upon them and her government is diligent in providing industries, investors, and businessmen with all the information it can make available.

The second government agency for economic control and planning is the Central Bank, located in downtown Amsterdam. The government views the main function of the Central Bank to be that of guardian of the guilder. It achieves this largely through monetary policies. The Central Bank also acts as the government comptroller office and the depository of domestic Dutch gold reserves.[8] The Central Bank is also responsible for supervision of the Netherlands banking system.

The Dutch have a long history of successful banking. Holland became the world's prime lender during the Golden Century of Dutch trade expansion in the 1600's. Though her banking monopoly has been eclipsed by the other lending capitals of the world – London, Paris, Frankfort, Basle, the large cities of America – she is still an important money lender in international trade. Today the Dutch banking system has been centralized and streamlined with the Central Bank acting as a clearing house and supervisor of all Dutch banking assets. There are approximately 100 commercial banks in the Netherlands, but three large banking units stand out. These are the Rabobank, Algemene Bank Nederland and the Amsterdam-Rotterdam Bank, familiarly known as Amro. All are products of a pattern of financial consolidation that has gone on continuously during the nation's post-war prosperity years. The Raiffeisen, a type of savings and loan, and the Boerenleenbank, a farmer's loan bank were the latest in this pattern, merging in 1971 to become the Rabobank, the nation's largest with assets approaching $ 8 billion. The Algemene Bank Nederland was an earlier merger of the Twentsche Bank and the Nederlandsche Handelmaatschappij; the Amsterdam-Rotterdam Bank, a union of consolidated banks of the country's two largest cities. There are also numerous private banks. There is also a small business loan bank, the Nederlandse Middenstandsbank, and an export-import bank, the Nederlandse Overzee Bank N.V. Important is the quasi-government special capital investment bank, the Nationale Investeringsbank with central offices in the Hague. The Investeringsbank was formerly the Herstelbank (Reconstruction Bank) responsible for the administration of the combined $ 1.1 billion Marshall Aid and matching $ 1.1 billion Dutch funds. The Investeringsbank, in addition to lending capital

[8] Dutch gold reserves stand at several billion, most of which are held in deposit in New York banks to offset foreign purchases. America is the one large nation with whom the Dutch have a heavy imbalance of trade.

funds for expansion, is also concerned with helping financially hard pressed Dutch enterprises. Other quasi-government banks include the Bank voor Nederlandse Gemeenten, primarily dealing with municipal loans, and the Rijkspostspaarbank, or postal savings, both of which are concerned with tapping savings resources for government capital investments. The Rijkspostspaarbank plays an important role in creating thrift habits among the youth of the Netherlands. All these banks fall under the aegis of the Central Bank.

The Central Bank handles the routine fiscal affairs of the Netherlands government. It processes all receipts and disbursements of government monies. This includes all social security, work and unemployment compensation and municipal accounts. As earlier stated, municipalities have little income of their own, and receive instead a pro-rata national revenue sharing that amounts to approximately 20 percent of the national budget. The Central Bank is also responsible for the money supply in circulation – that is coin and paper money for day to day transactions. This money supply is rather high amounting to about 23 percent of the total GNP. The Central Bank is also the clearing house for all bank transactions, similar to our own Federal Reserve Banks.

The Central Bank was created by the Banking Act of 1948 and the Credit Control Act of 1950, which added credit control to its functions. The Central Bank is run by a President but is controlled indirectly by the Council of Ministers, the Central Economic Commission, and the Social and Economic Council. Ministers of Finance, Economic Affairs, Social Affairs and others having to do with economic matters represent political control. The Central Economic Commission is composed of the top civil servants in important planning and control institutions. Both the Director of the Central Planning Bureau and the Central Bureau of Statistics are members. This body represents the professionals. The Social and Economic Council is the same advisory body that exercises influence over the Central Planning Bureau. The government is obliged to ask and listen to the advice of the experts – who thus have indirect power in government decision.

The Central Bank involves itself not merely with monetary policies but also with wages, prices and capital costs. Inflation in the economy is its present major concern. While it has no formal relationship with the Central Planning Bureau, the two organizations are in constant liaison, and the Central Bank eagerly awaits the Central Planning

Bureau's reports in order to make its own decisions. The Central Bank's major check upon the monetary flow of the Netherlands banks are its tri-annual directives. These establish the rate of loan expansion member banks are permitted by law. A high rate of loan expansion leads to ready inflation; too low a rate to possible business stagnation. This is similar to the dilemma of our own Federal Reserve rediscount rate levels. Presently the Central Bank is curtailing loan expansion to a one percent level during the latest reported four-month period.

Equally important is the role the Central Bank has in helping the economy maintain a balance of payment level. It shares this responsibility with the Central Planning Bureau and the Social and Economic Council. Despite the troubled international trade situation of 1969 and early 1970 the Dutch economy and balance of trade maintained healthy levels throughout 1970 to 1974. Dutch industry and trade were only marginally affected by the turmoils of the American market. There was no significant slowdown in Dutch economic growth – except for housing and government investments. There was a modest growth in industrial backlog, some rise in prices and a modest hike in wages. Workers were liable to be disappointed by the government's refusal to extend social security and holiday benefits.

Still throughout these years the Central Bank feared overheating of the economy due to growing demands in the public and private sector. *The Central Economic Plan* predicted a growth rate of 8 percent in 1970, but predictions for 1974 had receded to 5.5 percent.[9] Balance of trade showed modest deficits which could be easily adjusted by applying to the International Monetary Fund. Though not wishing to fight European inflation so strongly that it might precipitate an international slump, the Central Bank still recommended that the government restrict wage increases, decrease its own deficit spending, do less borrowing and expand its bond services. These measures were largely seconded by the Central Planning Bureau. One is impressed by the confidence and almost unanimity among the various "professional" government economic advisory institutions.

The most frustrating aspects of the Central Bank's efforts is that its control is limited largely to monetary policy. Policies adopted by the

[9] "The Netherlands Economy" *Economic Quarterly Review*, June 1970, No. 21; pp. 13–17, and J. J. H. Jacobs, "Aspects of Fiscal Annual Profits," *Economic Quarterly Review*, Amsterdam (Amsterdam-Rotterdam Bank N.V.), 1974.

government in expenditure, wage and price controls and taxes can often work counter to Central Bank goals. While government policies are usually "correct" on wage and price controls, its budgetary policies are often "inflationary financing." The government usually attempts to finance much of its expenditures by long-term loans which unfortunately are not fully subscribed. Short-term loans are resorted to, creating both deficits and easy money. Government loans are, in effect, "new money" which cancels out Central Bank restrictions on bank loan expansion. The Central Bank attempts to hold the inflationary line while the government, subject as it is to political pressures, overspends. This is particularly true in recent years. Political leaders find difficulty in resisting worker demands for higher wages, business demands for higher prices, and no one cares to offset this by unpopular taxes. Still the Dutch have proven themselves quite restrained. While inflation is a major problem, all is not black and there is little evidence as yet of "runaway inflation." A policy of strict price controls coupled with sluggish growth of domestic spending has kept it moderate. With the exception of the German Federal Republic, it remains the lowest in the OECD community. The guilder was so sound that there has been a substantial re-evaluation of it vis-à-vis the American dollar in 1972 and again in February 1973.

A third technical institution instrumental in the economic planning process in the Netherlands is the Central Bureau of Statistics. In its procedures and thoroughness we see once again Holland's reliance upon system and industry – "force of methods." The Central Bureau of Statistics' functions go much beyond merely serving the needs of the Central Planning Bureau and the Central Bank. The Central Bureau of Statistics offers its services to government, industry, trade, educational and research institutions, international bodies – even the mildly curious individual citizen or foreigner. The founding of the Central Bureau of Statistics came rather late in 1899. The history of its origins are peculiarly Dutch. The first general population statistics of the Netherlands took place under private auspices and were of limited value. In 1825, Lobotto, a civil servant of the Ministry of Home Affairs convinced the government of the need for an official yearbook which he published as his own enterprise until 1849. A statistical bureau was formed in 1826, published its first decennial census in 1829 and was disbanded in 1830. It was revived in 1848, and from 1851 to 1868 published an official yearbook. This second bureau was abolished in 1877. A Ministry of Finance was created in 1847 and it published its own statistics. Scientific, scholarly and business circles pressed for a national statistical center. A private association, Vereniging voor Statistiek, published a yearbook beginning in 1882. This organization was successful in creating a quasi-governmental body, the Central Commission for Statistics, which compiled, with government cooperation and sanction, monthly bulletins and annual yearbooks. In 1899 when the Central Bureau for Statistics was formally established, the Central Commission for Statistics became its control board. The former secretary of the Commission was appointed Director of the Central Bureau of Statistics. Until 1932, when it was shifted to the Ministry of Economic Affairs, the Central Bureau of Statistics operated under the Ministry of Home Affairs.

By law the 50-member Board of the Central Commission for Statistics is responsible for supervision of government statistics in the Netherlands and for control of the Central Bureau of Statistics. This pattern arrangement, with which we are now so familiar, serves to isolate and protect the professional operation of the Central Bureau of Statistics from the pressures and turmoils of the political processes. By virtue of the 1936 law "all citizens and other persons residing within the Kingdom in Europe, with the exception of heads of government departments, are obliged to supply the desired data and information."[10] The Central Bureau of Statistics on its part has an obligation not to use the data for any purpose other than statistical. The Central Bureau of Statistics has no regional offices. Most of its data comes from local municipalities (whose excellent records pre-date the Central Bureau of Statistics' existence) and from mailed inquiries and questionnaires. Other government organizations, such as the Central Bank and Custom Authorities, collect data of a limited nature for the Central Bureau of Statistics. Some inquiries, road traffic counts, social preference or behaviour questionnaires, consumer studies, etc., are made in the field.

The Central Bureau of Statistics has at its head a Director-General. He is aided by a deputy director, and an advisor in mathematical analysis, who is also responsible for data used in international reports. There are eleven statistical divisions, a division of National Accounts (which in close cooperation with the Central Planning Bureau analyzes the

[10] *Netherlands Central Bureau of Statistics, Organization, Function and Acts*, The Hague, p. 12.

significance of government incomes and expenditures), a department of statistical analysis and econometrics, a department of data processing, and finally a householding division. Approximately 70 percent of its employees are in statistical work, 15 percent in data processing, 14 percent in auxiliary capacities. Before computerization was introduced in 1960, 70 percent of personnel were engaged in clerical and routine tasks. The work of the Central Bureau of Statistics increasingly demands personnel of higher skills and training. Of the 100 university trained personnel, 70 are economists, the remainder mathematicians and statisticians. There has been a constant growth of the bureau matching the increasing reliance of government and private sector upon its functions. Its employees number close to 2,000. In 1974 the Central Bureau of Statistics underwent a reorganization to enhance its effectiveness.

A measure of its usefulness to government, industry and the public can be seen in the scope of its statistical reports. Two standard annual publications are the official *Statistical Yearbook* (*Jaarcijfers voor Nederland*) and the more digestible *Pocket Yearbook* (*Statistisch Zakboek*) which alone sells about 20,000 copies. Other regular publications include the *Monthly Bulletin of the Netherlands Central Bureau of Statistics* and the bi-weekly flyer sheets, *Statistical Bulletin*. These last give timely data in a uniform presentation that fits into the other Central Bureau of Statistics publications. Finally there are the Monthly Bulletins in specific areas such as agriculture, manufacturing, construction, foreign trade (by commodities and countries), transport (domestic, international, shipping, rail, canals, roads), finance, population and health, social statistics, police and justice. There are also technical publications on mathematics and models relating to statistical methods.

The Central Bureau of Statistics, in keeping with Netherlands' concern for international cooperation, contributes data to the United Nations Statistical Commission, the Economic Commission of Europe, the European Steel and Coal Community, the European Economic Community and such diverse organizations as the Benelux Economic Community and the Rhine Waterway Commission. Officials of the Bureau attended conferences among European and world-wide organizations.

We can more fully comprehend the Dutch passion for thoroughness of organization if we look into some of the work of the various statistical divisions of the Central Bureau of Statistics. The division covering trade is the largest in the Bureau. It covers data on foreign trade by commodity and country. This division handles over 30,000 forms daily, mostly from customs officials. It also covers almost all the domestic businesses, textiles, foodstuffs, processing, hotels and restaurants. It conducts comparative studies of productivity, enterprise costs and wholesale price indices. These are published on bi-weekly, monthly or annual basis and sold publicly. The division conducts studies for private business at a cost fee basis. It publishes a running account bi-weekly on key import and export items that have an avid demand. Statistics cover public finance, income, money supply, capital markets, property holdings and the distribution of wealth by geography and traditional socio-economic classes.

The Division of Social and Economic Statistics covers wholesale and retail prices in service and handicraft industries, consumer price levels and family budget studies.[11] It also covers labor statistics, employment, absenteeism, disputes, earning, salaries and figures of wages in industry, transport, agriculture, etc. A division is concerned with population figures relating to housing needs and economic recovery. In addition to its regular annual and monthly report, the Population Division makes frequent demographic forecasts. The most recently published report in 1973 made demographic projections into the year 2000.[12] This report has formed the basis for the many decisions taken by the Physical Planning Service in its predictions for the Netherlands' future in the new century.

It is evident that hardly a facet of the economy or society is overlooked. More important is the fact that this data is sought after and put to use. When one considers the economy of the Netherlands, one must acknowledge that he is dealing with a sophisticated society. Dutch social life may be rather uncomplicated, but its economy and technology are not.

How has economic planning fared in the Netherlands? On balance it is fair to say that planning has been successful. Largely under government direction, there has been a rapid post-war reconstruction. When the GNP of the Netherlands rose from $ 4 billion in 1945 to $ 65 billion

[11] These last are comparable to consumer studies carried on at the University of Michigan.

[12] Population Projection for the Netherlands for the Period After 1972, Central Bureau of Statistics, Government Publication Office, The Hague, 1973.

in 1974, wages and living standards rose as well. Real per capita income growth averaged about 3.5 percent annually, second only to that of the Federal Republic of Germany and well above the sluggish 1.5 percent for the United States. A favorable balance of trade was generally maintained and industrial productivity increased throughout this period. Planning was also responsible for altering much of the economic structure of the nation from pre-war years. There was a movement away from agricultural employment to the more profitable industrial and service areas, as well as deliberate effort, not always successful, of developing the less industrialized regions of the Northwest and Southeast.

However, what can be said of the immediate present, and the future, particularly since the problems of economic saturation, inflation, and world economic crisis have come to threaten the Dutch economy? Recent events in world finance have had their effect upon the Dutch economy and prompted their government to alter their plans and controls. Inflation seems to be endemic to the European and American markets. The first year of the seventies will go down as a year of unprecedented rises in wages and prices in the Netherlands. Wages rose 14 percent – well above the nine percent norm of the previous decade. Consumer prices rose seven percent, industrial costs from eight to nine percent, and housing, always in tight supply, an amazing 12 percent![13] There was a slowdown of the growth of industrial productivity in 1971, half the growth rate of a year earlier. This cutback was undoubtedly tied to the reduced disposable income that inflation sliced from the workers' wages and the continued high taxes affecting worker and employer alike. There was also a pruning of investments as businessmen, reacting to the sharp inflationary rise in costs, and tensions in the foreign investment market, adopted a "wait and see" attitude. Government spending remained high, particularly on housing and defense outlays. While there was a slowdown in total world trade, the Dutch share of the market increased by 3½ percent. Unemployment rose even in the volatile West, traditionally an area of high employment demand.

1972 saw a relaxation of tension in the Dutch economy with a stabilization of some of the inflationary indicators. Food costs averaged an eight percent increase. Though staples such as potatoes and some fruit showed little change, beef prices rose by 20 percent. Rents es-

[13] *Central Economic Plan*, Central Planning Bureau, The Hague, March 1972, p. 2.

calated by eight percent, a decline from the previous year's rate. Medical costs rose the most strongly, registering a 14 percent increase. Industrial costs held the same levels of increase of the previous year. There were continued low levels of investments in 1972 in all areas but building. Industrial investment improved slightly, but government investment declined sharply. Industrial recovery rose by 8½ percent, with the petroleum and chemical industries as the leaders, both with an increase of approximately 15 percent.[14] Unemployment came to quite a rise considering the half of one percent level that prevailed throughout the prior decade. In 1972 it stood at 100,000, or three percent of the work force.[15] Despite wage increases of 1972 and the re-evaluation of the Guilder in December of that year, the anticipated loss of trade resulting from increase in export prices did not occur. Netherlands trade held at six percent of the world's total while her neighbors, the Federal Republic of Germany, Great Britain, Italy, as well as Japan saw theirs decline.[16]

Throughout 1973 inflation continued to be, as it was for all major industrial nations, the major problem of the Dutch economy. The goal of the government was to break the cycle by gradually introducing wage restraint to halt the price spiral. The leveling of investment and industrial growth was seen as stabilization of the economy rather than a pessimistic portent of the future. The government exercised its tax regulator on incomes, the Value Added Tax and a tax on motor vehicles, to drain off monies that could fuel inflation. It curtailed spending, particularly on the Armed Forces. Unemployment became a growing concern in 1973. Proposals for cutting back the number of foreign workers, of widening the 40 hour week, and of lowering the pension age were seen as ways to reduce unemployment. Another area

[14] *Economic Quarterly Review*, Amsterdam, (Amsterdam-Rotterdam Bank N.V.) September 1972, No. 30, pp. 12–13.

[15] Interview, July 15, 1972, Dr. J. H. C. Lisman, Adviser, Centraal Planbureau, The Hague.

[16] "Setting the Stage for World Trade," New York Times, Financial Section, Dec. 31, 1972. Nothing better illustrates the reliance of the Netherlands upon foreign trade than comparative figures of exports and gross national product of that country with some of her leading trading competitors. (Percentage of world trade is given and the GNP is shown in parentheses.)

Netherlands	6.2%	($ 32 billion)	France	9.2%	($ 148 billion)
United States	18.2%	($ 991 billion)	Great Br.	7.7%	($ 121 billion)
West Germany	17.2%	($ 186 billion)	Italy	6.9%	($ 93 billion)
Japan	11.7%	($ 196 billion)	Belgium	5.5%	($ 26 billion)

receiving considerable attention by the economic planners was that of environmental policy. A program was advanced to reduce air, water and solid pollution to acceptable levels that would extend over a ten year period. Such actions cannot be affected without some direct costs upon almost every member of the society whether wage earner, stockholder or consumer. While our government refuses to spend half the $ 5 billion already appropriated for pollution control, viewing it as inflationary, the Dutch, with a much smaller economic base are prepared, unhesitatingly, to spend one quarter of a billion in one year for such a purpose.

Events of late 1973 and throughout 1974 have greatly altered Dutch plans to control their economy as they would have hoped. Their problems with inflation, unemployment and trade balances were vastly complicated by world events, particularly the oil crisis and the rise in food prices. Inflation pressures throughout 1974 spearheaded a wage demand increase of 15 percent which were countered by only a 10 percent increase in domestic prices. This proved of little advantage to the worker for higher wage and income taxes left him with no greater disposable income. Unemployment became more severe. At the end of June the total labor reserve, allowing for seasonable adjustments, stood at 144,000 or 3.5 percent of the work force. It was an increase of 12.2 percent from March of the same year.[17] Accustomed to an almost constant 1.5 percent unemployment these new levels must seem catastrophic to the Dutch. There is no way to imagine how they would react to the projected 8 percent unemployment rate for America in 1975. Surprisingly the Netherlands fared well in its 1974 balance of payments. Her surplus stood at $.9 billion, down slightly from the 1973 $ 1.2 billion figure. This was despite an alarming 30 to 35 percent increase in income costs that were met by only 25 to 30 percent increase in export prices. Undoubtedly much of this good showing is because her largest trading partner, West Germany, has the most stable economy in Europe.

In the September issue of the Central Planning Bureau's *Macro Economic Outlook 1975*[18] Dutch planners were still confident for 1975. They foresaw a favorable balance of trade increasing to possibly $ 1.2

[17] *Op. cit.*, Jacobs, p. 16.

[18] *Macro Economic Outlook 1975, Introduction and Summary*, Central Planning Bureau, The Hague, September, 1974. This informative publication provided much of the recent statistics for the final years under discussion.

billion. They forecast a mere 4 percent increase in import prices – a most optimistic contrast to her 30 to 35 percent range of 1974. They were equally confident that inflation would abate. Anticipating a slowdown in world trade, they saw, except for oil and food, declines in import prices. In the past three years prudent government budgets possibly contributed inflationary curbs upon the economy. However, the 20 percent increase projected for the 1975 budget may undo somewhat these salutary effects. Unemployment remains a Gordian knot.

Economic dilemmas persist. Dutch forecasters seem to be basing their projections on the assumption there will be no worldwide recession, that the West German economy will maintain its present strength and that domestic wages and prices can be controlled. The world economy is decidedly gloomy, undermined as it has been by rising unemployment, declining GNP's and trade disequilibrium, brought on largely by the financial crisis of a three-fold rise in Arabian oil prices. The effect of this last event has perhaps still to be fully felt in world markets. The wealthiest of industrial nations may well be faced with insolvency if this last maladjustment is not corrected. An imbalance of the order of \$ 40 to \$ 50 billion may well occur in 1975 to the OECD nations alone. World financial institutions are ill prepared to handle the gargantuan oil purchase loans or the recycling of petrodollars all this involves. Extraordinary measures such as the Federal Republic of West Germany's loan to shore up the Italian economy or the formation, with American cooperation, of a \$ 25 billion "pool" for the purchase of oil may prove to be only stopgaps. The Netherlands will undoubtedly pursue her traditional policy of international cooperation to seek out means to ameliorate present trade and monetary uncertainties. She will use whatever weight she possesses to oppose a return of rising tariffs, trade wars and monetary protectionism. In addition to the nagging "free float" of her currency the Netherlands has now to confront increasing costs of food imports, due largely to competitive demands on American grain and the temporary cutoff from Arabian oil imports.

Domestically inflation may not be brought under control and unemployment may continue to rise. The dilemma facing the Netherlands, similar to that of the United States, is that what would solve one problem would serve only to aggravate the other. Inflation can be curtailed by "cooling" the economy which invariably leads to higher unemployment. The unemployment cost may prove too high and the Dutch government may be forced to adopt a spending policy to offset it, only to unleash further inflation. There are clearly numerous uncertainties now present in the Dutch economic future.

As we conclude this survey of economic planning in the Netherlands, there are a number of observations that can be made. Despite the elaborate structure devised to permit expression of others, economic planning is centralized and largely controlled by the government. Yet the economic system of the Netherlands is not centrally directed in a hierarchical or monolithic manner. The Netherlands has a free economy. Industries, unions, and consumer groups have considerable autonomy. There is a long-standing tradition of individual economic orientation. Planning is above all, flexible. In the immediate desperation of the post-war period the government had much to say but the later post-war thrust was towards a "normalization" of a free, capitalistic economy. However, given the particulars of the Netherlands economy – high level of industrialization, dependency upon world markets and demands for equitable distribution of wealth – can a totally free economy be resorted to?

Almost every facet of Dutch society plays a part in determining economic plan policies.[19] The public voice is heard in the election of Parliament. The Government (Cabinet) is a result of a majority coalition of parties that can agree upon a common policy. The Economic Affairs Council, those Ministers whose functions are largely economic, plus exofficio members from the Central Planning Bureau, Central Bank, Social and Economic Council, etc., determine economic planning policy for the Cabinet. They in turn are advised in their decisions by the non-governmental organization – the Central Economic Committee. Two ministries, Economic Affairs and Finance, are directly responsive to the advice from the Central Planning Board and Central Bank, respectively. The operations of the Bureau of Statistics are made available to these last two. The quasi-government Social and Economic Council is responsible for recommendations to the government on general conomic policy. The Central Planning Bureau and the Bureau of Statistics are permitted to operate in a wholly professional manner, freely criticizing the government when in its judgement the government has erred. The Central Planning Bureau in turn calls upon business, labor, educational and research institutions, even international organizations for information, advice and opinion. The ultimate "plan"

[19] See chart page 44.

arrived at tends to be suggestions, explanations rather than mandates. The government retains power and assumes responsibility. It has a corps of competent professionals as advisors. All points of view are expressed before final decisions are taken. Planning in the Netherlands is organized, thorough, flexible, democratic – and one might add totally Dutch.

It is possible to say that economic planning in the Netherlands has been successful, if measured by increased industrialization, growth of population and productivity, raised standards of living, and more equitable distribution of wealth. It is more difficult to gauge to what extent this success is attributable to economic planning or to the general improvement of world economies. Present economic events and their future outcome will be enlightening.

When the American system of economic planning and control is compared with that of the Netherlands it seems to be no system at all. There is much more government power to direct and control in the Dutch system. There are not merely the traditional devices of taxation and government spending levels, the fiscal measures of mandatory bank reserves and discount rates, but the Netherlands government has also, while it prefers cooperation, direct control over wages and prices. In fact the government has legislative power to control almost any facet of the nation's economy. Moreover, the Netherlands government has an organized structure to implement these control measures. Important is the fact that government planning and control institutions have earned the respect and support of the Dutch public. The level of professionalism in the Netherlands government is higher than in the United States. One often suspects American government reports and statistics of being put to political uses. Statistics on prices, the cost of living index, and unemployment figures often seem contradictory. The American government has not developed the technique for forecasting, planning, or control that matches that of the Netherlands.

There are, of course, reasons for this. The most important is that of size and manageability. The dimensions of the American economy make almost any attempt to organize and control it seem an exercise in futility. We have an entropic dilemma that even computerization seems unable to master. Certainly, thirteen million people with an economy producing $ 65 billion of wealth is much less a tangled labyrinth than a nation of 210 million with over a trillion dollar economy. Moreover the impulse to planning is more pointed in the Dutch economy. America does not rely upon foreign markets for over half its income.

There remains one further consideration, that of popular attitudes towards the role of the government in a nation's economy. The Netherlands has experienced far more than America the rewards of economic direction and cooperation. Her geography and history have taught her different lessons. Literally, to keep his head above the water the Dutchman has learned to accept controls from above and to cooperate with his neighbor. He carried this over to his economy, organizing his efforts with others to have a competitive advantage in world trade. Both efforts have been successful. The industrial revolution came late to Holland and its dislocations were met with a variety of social legislation unacceptable to an America which underwent this process of industrial change completely captive of the doctrine of classical capitalism and laissez-faire. Even the lessons of the Depression and the New Deal achievements did not bring our society that closer to economic planning and regulation. Our insistence upon a completely unfettered economy is reflected in America's obsession with anti-communism, which has no true counterpart in the Netherlands. To us the economy remains sacrosanct – not to be tampered with. Perhaps it is this mentality, or social mind set, that prevents America from moving towards planning, control and redevelopment of her economic resources. We just don't think that way.

CHAPTER IV

THE PATROON SYSTEM – MODERN STYLE

Holland today is one of the most representative welfare states. In part this is a result of her population density, her industrial complexity, and her vulnerability to world economic fluctuations. Such a society cannot tolerate economic chaos or social dislocations. Holland's neat and thorough system of social legislation partially reflects her history of public charity that goes back to the Golden Century. This tradition has made social policies acceptable to her people that would be incomprehensible to the American whose fierce attachment to individualism and the doctrine of self-help exclude such notions of social responsibility towards the poor and unfortunate. Centuries of building dykes and fighting the seas also gave the Dutch a sense of social cohesiveness somewhat lacking in American society.

The immediate, and most important reason for the Netherlands' advanced social policies undoubtedly stems from the experiences of common hardships and humiliations of the Nazi occupation of their land. Out of it came a sense of social solidarity and a determination for social and economic equity for all. The Dutch people have forced their government to become a servant to this wish. It was a wish to which the government and political leaders readily acceded. Rarely has there been a nation in modern times so confident of its social policy. The present does represent a period of hesitancy, but almost all elements of the Dutch society view the immediate past with great self-satisfaction.

Compassion is a part of the Dutch tradition. Calvinist charity and bourgeois "noblesse oblige" brought about a high level of social concern from the beginnings of the Republic. Its "Begijnhofjes" were models for their day. Hospitals and orphanages existed in every major city. In the judgement of one of the foremost contemporary historians of Holland, "no society provided so amply for its poor as did the governing merchant class of the Dutch Republic."[1] The House of Orange maintained a paternalism that from early times captured the affection and loyalty of the people.

Royal concern has been sincere rather than purely expedient. Willem I, the so-called "Merchant King" exhibited a great energy in forwarding the economic development of his country. Royal initiative was behind such measures as the establishing of the textile industry in the East, the North Sea Canal project, and the expansion of the railroad system. The textile industries saved the people of Drente from destitution. The canal project was undertaken as much to provide poor relief as to provide Amsterdam with access to the sea after the silting up of her Zuyderzee approaches. Railroad building was similarly an aid to overcoming national unemployment. However, the tradition of royal assistance did disincline the people from exercising their own solutions to their problems.

The Industrial Revolution came late to the Netherlands. Her reliance upon an economy of agriculture, shipping and trade, postponed her industrial development until the middle of the 19th century. Despite this, when the Industrial Revolution did occur it produced the same pattern of social displacement and misery that happened earlier in England. The suffering of the Dutch working class was unbearable. Extended families were herded into single room hovels. Diets barely sustained life. Hunger and disease were widespread. Tuberculosis was in the very pattern of existence. Unemployment was so rampant that even exploitive employers of women and children were looked upon as benefactors. Working conditions were unimaginable. Crime, prostitution and drunkenness were part of the prevalent morality. The poor lived in rags without sustenance or hope. Society stood at the point of collapse. Chronically 16 percent to 20 percent of the population was on poor relief.[2]

The Dutch lower class had a habit of docility born of faith in the traditional benevolence of their City Fathers and the patronage of the Crown. These and religiously sponsored charity, had been the cure for social misfortune in ages past. But now, the Catholic and Calvinist churches, in both their moralities and efforts, seemed irrelevant. The ameliorative measures of the liberal middle class were well intentioned

[1] Wilson, Charles H., *The Dutch Republic*, New York, (McGraw Hill) 1968, p. 53. Used with permission of McGraw-Hill Book Company.

[2] Kraemer, *Op. Cit.*, p. 34.

but ineffective. There had been no experience in social responsibility that prepared Western society to cope with the economic and social maladjustments of the Industrial Revolution. Holland's history was no different. Despite their plight the working masses of the Netherlands did not join in the open rebellions of 1848. Still the people were not left totally unaffected by the social utopianism and labor socialism of that era. Class consciousness and democratic aspirations had been awakenend earlier with Holland's flirtations with French revolutionary concepts during the time of the Batavian Republic. The constitutional reform granted in 1848 was received by a people ready to revise their attitudes and expectations concerning their government and their social order.

The initial supremacy of the Dutch legislature coincided with the liberal leadership of Thorbecke. As the society was breaking away from the fatalistic view of society as a God ordained order wherein the rich were predestined to be rich and the poor to be poor, political leadership was turning its attention to the "social question" – stagnation of the Dutch economy, poverty, and economic injustice. In 1854 the Poor laws were passed permitting local governments to rescue church and private charities. Educational reforms were introduced. The franchise was broadened. Religious and political disabilities against the Catholic minority were partially removed. As early as 1851 the Municipal Corporations Act gave cities the power to appropriate private property for roads and canals. Eventually these powers were extended to public housing and city planning. In 1875 the Public and Private Nuisance Act gave government the power to interfere in industry when personal or public health was threatened. A Housing Act was passed in 1901 that was a model for its time.

The legislature continued the tradition of economic responsibility so wisely set by the Crown. To enhance trade and to decrease unemployment the government sponsored the building of the Calland Canal and the enlargement of the Rotterdam port. It did the same for Amsterdam when it constructed the IJ Canal and the world's largest shiplocks at IJmuiden. Its early 20th century efforts to increase economic development and to decrease unemployment made the national government the largest single partner and sustainer of private industry. The government took over the unprofitable coal mines in the Limburg region rather than countenance the massive unemployment that their abandonment would entail. It became the majority stockholder of KLM through its effort to promote Dutch commerical aviation. It became the promoter and later major owner of the new steel mill complex, Hoogovens, at the mouth of the IJ. It granted large subsidies to other developing industries. It had long since taken over control of the railroads, telephones and telegraphs. The state had permitted municipal governments to become corporations in their own right, particularly in utilities, public transportation and housing. The Zuyderzee Reclamation project and the ambitious Delta Plan are present-day illustrations of the pattern of government responsibility for economic well-being that is now so typically Dutch.

Much of the impetus for this change came from the efforts of the working poor. The Industrial Revolution posed problems that seemingly only a labor movement was able to solve. There were the contributions of liberal bourgeois conscience - Dr. Aletta Jacobs' efforts for child labor legislature, public housing and public health - and of capitalist paternalism as exhibited by the work of S. Le Poole, J. C. Van Marken and C. T. Stork. Labor socialism had late beginnings in Holland. The first nationwide labor organization, the Algemene Nederlandse Werknemers Vereniging, was not created until after 1850. The first socialist elected to Parliament, Domela Nieuwenhuis, took office in 1888. A socialist association, which quickly moved to trade unionism on the English model was established in 1871 as the Algemeen Nederlands Werkliedenverbond. The temporary improvement of economic conditions following 1870 helped to make reform easier. The call for social reform urged by Rerum Novarum in 1891 impelled many Catholic worker organizations to ally with the socialists for chance. Yet only by 1894 were Dutch workers able to accept a Marxist Socialist Party, headed by Troelstra. Pressures for labor and social legislation grew apace.

In 1874 Van Houten had been able to push through Parliament the first child labor law. This measure to prevent employment of children under twelve years of age had been asked for by a committee of 32 manufacturers ten years earlier! In 1889 legislation limited the employment of women and children to a merciful eleven hours daily. Not until 1911, following the Berne Convention of 1906, did Dutch legislation limit their working hours to ten. The more acceptable eight hour day came in 1920, after nine years of deliberation in Parliament. The five day, 40 hour week is still not totally in effect throughout the Netherlands.

There was reluctance to move rapidly towards satisfaction of labor demands on hours, working conditions, and social insurance. In 1895 the first legislation of working conditions detailing light, ventilation, protection standards – and the reporting of accidents was passed. By 1900 such measures were fully acceptable. The first worker social security legislation came in 1901, the same year that saw the passage of the Workers' Housing Act. Added later was the Industrial Accident Insurance Act of 1913 (Ongevallenwet) which gave limited workmen's compensation to certain industrial occupations. This was enlarged by subsequent acts such as the Seamen's Accident Insurance Act (1919) and the Agricultural and Horticultural Accident Insurance Act (1921) which specifically covered other occupations previously excluded. The Industrial Accident Insurance Act was amended in 1921 to extend coverage and benefits. A Sickness Insurance Act was also passed in 1913. In 1919 another landmark was achieved with the passage of the Invalidity Act, which provided life pensions for widows and pensions for orphans to their majority. The reluctance of society and government to travel this road can be seen by the delay of six years between its passage and implementation.

The hardships caused by Holland's isolation during World War I, the Depression of the 30's and military occupation gave impetus to social legislation. In 1930 the 1913 Sickness Insurance Act (Ziektewet) was expanded to cover not only sickness benefits but medical treatment and even maternity costs. The improvement and control of sick fund management and the Children's Allowance Act (Kinderbijslag) came under government review ending the earlier 1941 decrees of the occupation power. Earlier in 1939 the government took upon itself to be the arbiter of all labor questions. Collective bargaining was assured by measures passed in 1923 and 1927. Government employment offices were created in 1930. In 1939 the government assumed partial responsibility for unemployment compensation. The lateness for this measure can be properly judged only when it is understood that traditionally it was the unions that had the role of alleviating labor problems. The Act of 1939 followed the Ghent Plan, with government matching equally the contributions of unions to unemployment benefits. In 1952 the government's role in unemployment insurance became total. In 1957 a General Old Age Insurance Act was passed.

All social security and worker benefits were reorganized and brought up to date in 1967. Additional amendments enlarged the scope of insurance coverage and benefits – the latter to match inflationary rises in the cost of living. Dutch social security is divided into two categories: Worker's Insurance and National Social Insurance. The first includes Health Insurance, a new provision covering short-term medical treatment costs; the Workingman's Incapacity Act covering long-term medical costs and pensions for invalids; unemployment benefits (Werkloosheidswet 1952); and the Wage and Earners Children's Allowance Act. National Social Insurance includes the Old Age Pension Insurance (1957); Widows' and Orphans' Insurance (1919); Children's Allowance Payments (1941); and General Sickness Insurance Funds (1930).

Worker protection in the Netherlands follows the pattern of all modern industrial societies. There are regulations providing for safeguards against the hazards of industrial employment. Working conditions, hours and wages, labor disputes, workmen's compensation, unionization, savings plans, workers' housing and even child allowance are included. Labor insurance under the Reorganization Act of 1967 covers health, accident, unemployment, child allowance and old age benefits. Health and accident are paid by the National Health Inusrance Fund. Sickness payments to workers amount to 80 percent of their normal pay up to a maximum of 400 guilders weekly, which approximates $ 156.00.[3] Pregnancies and confinements are covered for women workers who receive full payment during maternity leave. Accident insurance, like the health program, is compulsory for all workers. Benefits pay for all medical costs and disability pension in the event of work invalidism. A full disability is usually awarded a pension of 80 percent of normal wage. This can be extended to 100 percent if invalidism involves continuing medical attention and nursing care. The first year of medical and hospital care is assumed by the Health Insurance Fund provisions. There are appeal boards to which a claimant can repair for revisions or adjudications of his case. Approximately 1.3 percent of workers' wage covers premium for this insurance coverage.

Unemployment is af ull government responsibility. The normal pattern is for a released worker to receive severance pay for the first

[3] Amounts throughout this discussion are "rounded out" approximations. The rather significant differences in purchasing power of the two currencies make it difficult to equate true exchange value on the basis of official exchange rates. Another complication is introduced by the recent dollar devaluations. The official rate of exchange quoted in *The New York Times*, February 7, 1974, gave the guilder a value of 39.12 cents. It has eroded slightly, two years later being approximately 38 cents.

forty days of unemployment. This is paid for by the Industrial Association to which his employer belongs. After this period the unemployed worker is entitled to unemployment benefits amounting to 80 percent of his wage for the next 130 days. Premium costs average about one percent of workers' wage income. Unemployment payments were consistently below premium contributions, but recent high unemployment levels are taxing the national unemployment fund.

Workers are also covered by pension provisions of the Old Age Pension Fund Act of 1957. Corporations, and sometimes unions, have supplemental pensions, life insurance and savings programs for workers. There are over 1650 such independent funds which are carefully regulated by the government. Workers who do not agree in principle with life insurance plans are not forced to join them even if provided by the place of their employment. Employers are compelled, however, to place an equivalent amount of the premiums they would have paid under these plans into a savings program for the benefit of their employees.

The government, through its National Social Insurance program, also contributes greatly to the national goal of individual security for all its citizens. This is seen in its comprehensive and rather generous Old Age Pension, Widow, Orphans, and Family Allowance programs. The benefits of the Old Age Pension scheme go automatically to all citizens who have attained 65 years of age. The program is compulsory and payroll deductions are levied against all wage earners and the self-employed. No contributions are assessed beyond a certain income level. The income of a married worker is fully taxed up to a level where only a partial tax is paid, to a point where finally no further income is taxed. This principle applies to the single worker as well except allowances are not as generous. Benefits are substantial, approaching $ 200 per month for the single retiree and $ 290 per month for retired couples. This compares favorably with the average Old Age benefits of approximately $ 187 and $ 260 per month to similar recipients under the American Social Security system. This becomes more true when differences in living standards between the two countries are taken into consideration. Lower paid recipients of American Social Security benefits fall far below Dutch levels.

More significant is the fact that Dutch retirees receive equal pension benefits regardless of their contributions to the program. There are set pensions each for retired individuals and couples. The Dutch system tends to create an equal minimum standard of living for the aged. This also serves to make administration of the program far less complicated than in the United States. The American system perpetuates the inequities of income maldistribution by gearing its pension benefits to the worker's lifetime earnings. The wealthy who have lived more abundantly during their working years and were able to build a cushion of savings and property for a more secure and comfortable retirement receive the highest benefits. The poor who have lived on a reduced scale throughout their working lives are, because of lower Old Age benefits, penalized in retirement as well. Often the rich find government pensions not vital to their retirement, while to the poor, however inadequate they may be, pensions are all they have.[4] The equalization of pension benefits is an innovation that we might borrow to ameliorate somewhat the economic inequities that seem so self-perpetuating in our American system.

The Dutch government has apparently decided not to force its citizens to be left entirely upon their own resources in securing their future. It is ambitiously hopeful of extending benefits beyond the goal of scaling Old Age Pensions to match the alarming inflationary costs of living rises that are so devastating to those on fixed incomes. Increases in pension benefits in the Netherlands doubled from 1969 to 1973, with further increases in 1974. There is a provision to give the retired an additional 6 percent sum to equal the vacation pay allowance given workers in industry. American increases have not been as generous. On the average, recent social security benefit payments increased by only 30 percent. There is some concern that, expanding social security benefits as they are, the Dutch may well be straining their system beyond the limits of actuarial prudence. Still it seems certain that,

[4] The inequities of the American Social Security system is best seen in the minimum payment levels. A worker retiring at age 65 could receive as little as $ 93.80 per month. If retired at age 60, benefits would be a mere $ 75.10 per month. A widow at age 60 could receive even less. These all represent low income earners probably highly dependent upon government pensions for their security. These figures help to explain why one-fourth of Social Security recipients are on welfare. Under the present Social Security schedule the highest benefits to a single worker is $ 304.90 per month. A married couple's pension could range from approximately $ 110 to $ 450 per month depending upon their average incomes. The gap between higher and lower benefits has been narrowed, and the pension levels have been greatly raised in recent years. 1974 saw an 11% increase in benefits alone. Information Office, Social Security Administration, U.S. Department of Health, Education and Welfare.

despite her economic limitations, the Netherlands is committed to underwrite this program for a more socially equitable wealth distribution. Their program of concern for the aged, hardly ever discussed or remarked upon, seems taken for granted as it should be in any rational society.

Supplementing the Old Age Pensions are the Widow and Orphan benefits. The Widows' and Orphans' Insurance Act of 1919 was updated in 1959. It provides for immediate temporary payments, or permanent pensions to widows and orphans. If young widows are incapacitated or mothers of minor children, they receive pension benefits. Those over forty receive pensions regardless. Pension in 1973 for childless widows was approximately $ 185, while widows with children received about $ 256. Orphans also received benefits related to their anticipated living expenses. Monthly pensions are approximately $ 54 for a child under ten, $ 80 from ten to fifteen and $ 105 for children above that up to age 27, if they are still pursuing their education. The cost is 1.4 percent of the worker's wage. The fund is administered by the Social Insurance Bank, which handles all the social insurance arrangements.

The Family Allowance Act (1941) was modified in 1963. It provides for allowances to parents beginning with the third child and thereafter, provided they are under 15 and considered "a part of the family." There are additional allowances for children that are in any way incapacitated. These family allowances are distinct from those provided for under the Wage Earners Child Allowance Act, which are tied into the workingman's wages. Benefits under the Family Allowance Act begin with the first child and offer somewhat higher benefits. Both these acts are controversial. Holland's birth rates were, until just recently, among the highest in Europe, as is her population density. There are those who question the advisability of such programs that invite larger families when the nation is faced with the dilemma of population pressure on limited geography.

The Dutch system of social security and worker's benefits, like those of the United States, follows the pattern found in other modern, industrialized societies. There are, however, some important differences between the Dutch system and our own that need to be underscored. While methods of worker contribution to social security programs are similar, payment levels of distribution patterns are different. We offer, except through our Aid to Dependent Children programs, no direct children allowances benefits. These payments to unwed mothers and female heads of broken homes take a significant "bite" out of our total social security funds. The Dutch pattern of family life in general is highly stable, and the eroded family so frequently encountered in America is not, as yet, a significant problem to the Dutch. Regrettably, under our Aid to Dependent Children program, male heads of families are encouraged to desert in order to entitle the wife to subsistence and child support benefits. It is doubtful, however, with our increasingly higher levels of family instability, that direct children allowances employed by the Dutch would be helpful.

Included in the National Social Insurance Program are provisions covering injury and ill health. The National Health Insurance coverage for accident and sickness was extended in 1968 by the General Special Sickness Expenses Insurance Act. (Algemene Wet Bijzondere Ziektekosten). It is a relatively new social security benefit covering exceptional sicknesses or invalidism such as blindness, deafness, mental deficiency and mental illness.

The National Health Insurance Fund Act (1966) provides medical coverage for all Dutch citizens. Participation is compulsory for wage earners and voluntary for non-wage earners whose incomes are below a minimum income level. Prior to this act, health insurance policies offered by private insurance companies was the only coverage for this risk. Usually low income families subscribe through labor, company, professional or church organizations. There are over 100 such funds now regulated by the Ministry of Social Affairs and Public Health. Direct control is given to a quasi-governmental body, the Health Insurance Fund Council (Ziekenfondsraad) composed of 35 crown appointees, representing government, employer, employee and professional interests in equal proportion.

The Health Insurance Fund Act of 1966 is not to be confused with the Health Insurance Act which gives payment benefits to employees for losses due to sickness. Its payments are strictly for medical and hospital expenses. As a rule the Funds do not give cash benefits. Instead they pay directly the costs of medical treatment, hospital care, medical supplies, and doctor's fees. Often included are dental, maternal and mental health care, speech, hearing and physical therapy, prosthetic devices, X-ray and ambulance services. In this regard this corresponds to our Hospital and Medical Plans such as Blue Cross or Blue Shield. The individual is free to choose his own doctor and his own hospital. He is limited in his choice of specialist for this is usually decided upon by his physician. The patient can change doctors, but it does require making the change known to health insurance administrators.

Costs to the insured are reasonable. They are paid for by wage deductions, employer contributions or by assessment. The rate is near seven percent of the worker's wage up to a maximum, which in 1973 was, $ 100 per week. Half of this is being paid by the employer. Considering that coverage includes wives and children it cannot be deemed burdensome. Those already receiving other social insurance benefits – old age pensions, accident insurance, unemployment compensation, widow or orphan allowances – pay only half that premium. Persons not compulsorily insured may seek coverage under private health insurance funds. Private insurance companies are obliged to insure the aged at low premium rates geared to their lower incomes. Individuals can join Health Funds through corporation, trade union, church and other voluntary association plans.

Though medical costs have risen, there is not yet the runaway inflation of medical costs that are undermining public medical-health programs in the United States. Medical fees are set by the Sick Fund Board. The Dutch would find our method of regulation over Medicare and Medicaid program fees incomprehensible – comparable to placing a fox guardian of a hen house. Permitting the medical profession to be the sole regulator of its standards and costs has not always been to the public's advantage. The Dutch physician is paid on a pro-rata basis by the insurance funds. They receive a per annum rate for each of the first 2000 patients assigned to them, treated or not, under the plan. For those over this number their payment per patient is reduced. Thus a doctor with 2000 Ziekenfonds patients would have an income of about $ 17,000. A Dutch surgeon working with the advantage of hospital facilities would average $ 120 per working day. Dutch doctors receive at least $ 5.00 for a normal office visit outside their case load. Home visitations run higher, especially when they involve long distances or inconvenient hours. Dutch physicians *do* make house calls.

Medical rates, like all other costs, are rising, but still keep more in harmony with the inflationary rises than our own medical fees have proved to have been. There are also attempts by the Sick Funds to control "doctoring" and extension of medical care. Payments on case loads that extend over long periods are reduced, often halved. Physicians are encouraged to complete treatment so that they can be free to open up newer cases on full payment levels. Renewal cases are also reviewed by the Sick Fund administrators and payments on frequent "repeaters" are also lowered. Thus these are efforts to discourage extended or useless visits that cause increased costs to the programs.

Physicians are permitted private practice as well, but they seem content. While not apt to acquire fantastic fortunes they have a leisurely style of living. They are rather well-read and unusually well-informed. Hospital costs are also reasonable. Nor are hospital personnel the recipients of the rather depressed incomes so typical of our own hospital workers. There is a serious shortage of nurses presently met by foreign nurses from abroad, and the overseas Dutch territories. Better treatment of student nurses, higher pay and the creation of a more attractive public image might serve to enhance nursing as a career in the Netherlands. Medical standards can be rated as outstanding. This combination of professional competence and economic restraint is rarely encountered in societies not attuned to the philosophy of socialized medicine. Though there is some dissatisfaction among doctors, the cooperation of the medical profession with the nation's health program remains the key factor in its success.

The Dutch also have a highly developed public health program. Their public health arrangements follow the familiar pattern of government control and public participation. While the government task in public health seems to be that of setting standards and inspecting facilities, and assisting with financial aid, the bulk of public health work is carried out by private, voluntary associations. Government responsiblity comes under the Ministry of Public Health. Most of the technical aspects of the work is the responsibility of the Director General of Public Health and the State Supervisory Public Health Service. There are countless miscellaneous Public Health agencies dealing with specific public health problems – the National Institute of Public Health, the National Institute of Preventative Medicine, the National Institute of Pharmaco-therapeutic Research, the Health Research Organization (a component of TNO – the Institute of Technical Research in Delft), the National Institute of Water Supply (RID) and the National Institute of Waste Water Purification (RIZA). There are the inevitable advisory boards comprised of government and non-government members that is so much the style of the Dutch way of controlling their social institutions. These include the Public Health Council, the Central Council of Public Health, the Nutritional Council, the Rehabilitation Council, the Health Insurance Council, and the Council for Refuse Disposal – all elements in a typical modern bureaucratic complex.

Working public health facilities – hospitals, clinics, sanitaria – are largely provided for by the municipalities. The Factory Inspectorate takes the responsibility for industrial accident prevention and safeguarding against industrial health hazards. Many health services, routine dental care for example, are undertaken by schools. Health information is made public through state agencies. Crusades are carried out for immunization against contagious diseases when public health authorities deem necessary. There are routine immunization programs for typical children's diseases – measles, mumps, polio. Still much of public health activity is carried out by voluntary, private charitable associations. Of the 266 hospitals in the Netherlands, only one-fourth are municipally or provincially maintained. Municipal hospitals are usually larger, better equipped and do account for more than one-fourth of the nation's 71,000 hospital beds.[5]

Voluntary associations play the key role in non-industrial public health. They have assumed the primary responsibility for nation-wide maternal services, child care, mental health, medical therapy and tuberculosis sanitaria. In addition they maintain most of the nation's hospitals, clinics and other medical service agencies. These include baby clinics, child welfare centers and summer camps for children. They also maintain marriage counsel and psychiatric and rehabilitation centers for alcoholics. This last is a problem that has little gravity in Dutch society. There are over 22,000 mental patients also taking hospital space. The Dutch medical authorities and the voluntary societies have partially solved the problem of hospital space by treating as many cases as possible on an out-patient basis. In all this there is an almost inextricable web of private and public cooperation. Voluntary associations work hand in hand with government authorities with maternal and child care, with control of communicable diseases, with mental health, with rehabilitation, with public health information. They are liberally helped with public financial assistance but they do assume much of the financial burden themselves.

[5] *Kingdom of the Netherlands, Facts and Figures*, 1970–1971; (Government Printing Office) The Hague, p. 10, and *Statistisch Zakboek*, p. 45.

Voluntary societies also assume most of the task of caring for the nation's aged. In a society where about fourteen percent of the population is beyond 65 this is no slight burden. Hospital space for the aging is one of the serious public health problems remaining in the Netherlands.[6] Compassion for the sick and aged is seen as a social obligation. Of key importance is the countless hours of unpaid assistance given by public-spirited women and young volunteers. There is a noticeable level of charity and concern in the relationships between these workers and their patients. There is also a warmth and communication that seem lacking in our generally larger and more impersonal city hospitals. Dutch hospitals tend to be more homey. They are generally smaller, usually two stories, with grounds, walks, shrubs, trees, gardens. There are many older hospitals, particularly in older city centers. Though hospitals and homes for the aged tend to be crowded, they are generally not depressing. For one, they are usually well peopled with visitors, even children, who are normally not restricted from visits. There is more of a hubbub and much talk and laughter. The Dutch accept sickness and old age as facts of life and don't adopt the embarrassed and sanctimonious quiet Americans seem to in such circumstances. They are also more thoughtful of their aged and less apt to neglect them.

The voluntary associations are largely outgrowths of church-related social action. Denominations maintain their own hospitals, clinics and some, even their own medical colleges. Many nursing associations are also related to the churches. The Roman Catholic Church has its White and Yellow Cross; the combined Protestant Church, its Orange and Green Cross. Frequently these groups work with each other in addition to the Red Cross (Rode Kruis) which is viewed as being international in character. The Dutch also have their own non-denominational nationwide organization, the Green Cross.

A description of Green Cross functions reveals much that is typical of the others as well. For a minimum contribution of about $ 10 one can become a member and is entitled to a comprehensive array of cost-free services. A nurse or an aid will come to the home of a hospitalized

[6] Fortunately the authorities have been able to commandeer the abandoned tuberculosis sanitaria of the early post-war years. Due largely to the malnourishment of the Occupation years, tuberculosis had a high incidence in the fifties. TB has been virtually wiped out and over 8000 hospital beds have been released to ameliorate the problem of space for aged patients.

mother, or during delivery, to take over the task of caring for the young, preparing meals, laundering, etc. The Green Cross maintains clinics where courses in baby care are given to expectant mothers. They have scheduled patrols with baby buses that provide post-natal care for infants. Younger children can go to Green Cross clinics for sun-lamp treatments, no small service in a country notorious for its lack of winter sun. The handicapped are given therapy and prosthetic training. The aged are cared for in their homes, rather than being immediately hospitalized. All manner of equipment, special beds, wheelchairs, crutches, prosthetics are loaned to members in need. The Green Cross employs doctors, specialists, geriatricians, nurses, nurses' aids and domestic help. These cooperate with public health officials and welfare workers in making the life of the sick and aged more pleasant. There are over one and three-quarter million memberships in the Green Cross. In addition there are over one and one-third million members in the Catholic and Protestant voluntary Cross associations. Such figures give us a measure of Dutch concern for their unfortunates. The Dutch are not ungenerous to their children, sick, and aged. Most important is that they make them feel that they belong, and are not just simply in the way.

Medical standards in the Netherlands are consistent with the high level of technological advance of all of Northwest Europe. Medical training in the Netherlands is thorough and of high scientific calibre. There are eight medical faculties in the Netherlands with an enrollment of about 12,000, 20 percent of which are women. There is a medical faculty at the state universities of Leiden, Groningen, Utrecht and Amsterdam. There is a medical school at the Catholic university at Nijmegen and at the Protestant Free University at Amsterdam. There is also a medical school for clinical training at Rotterdam. Schooling for the medical profession takes from seven to eight years. Specialist training can add four to eight more years. Thus a Dutch physician has from seven to sixteen years of training depending upon his level of skill and study. There are about 600 to 700 medical school graduates annually for a population of thirteen million. To maintain that ratio, America would have to graduate more than twice its present number. The medical profession in the Netherlands has no quota system, though entrance qualifications are high. While there are shortages of medical personnel in parts of the Netherlands, there is not the total reliance upon imported medical professionals that keep America's public health institutions from total collapse. There is a shortage of dental personnel, and an impending surplus of pharmacologists. While women do not seem to move into dentistry, they do make up about one-third of student enrollment in the schools of pharmacology.

There are schools of midwifery at Amsterdam, Rotterdam and one under Catholic auspices at Heerlen. The first male midwives were trained at Heerlen. Considering that over one-third of the deliveries in Holland is accomplished by midwives, such training is important. Only one-third of child deliveries is made in hospitals. In the remaining third of the cases the mother is attended by a general practitioner at home. Still the infant mortality rate in the Netherlands is at the low figure of less than 17 per 1000 live births, compared to the United States statistics of 25.3 per 1000.

There is a total of 15,000 medical doctors in the Netherlands, divided almost equally between general practitioners and specialists. This ratio of 1.25 per 1000 population is slightly below America's 1.65 per 1000.[7] As in America, medical doctors tend to be concentrated in urban areas but possibly not to the degree found with us. Moreover population and living standards are more equally distributed throughout the Netherlands. Holland has no problem of an almost total doctor shortage that one encounters in rural North Dakota, Mississippi or South Carolina.

Dutch medicine is competent. We have already referred to infant mortality rates but other statistics are inclined in favor of Dutch medicine. Rates of death from infectious or communicable diseases are 0.9 per 1000 in the Netherlands compared to 1.2 per 1000 in America. The life expectancy span for the American male is 67, for a woman 75. Comparative Dutch figures are 71 and 76 plus. True, longevity may be a factor in Dutch heredity, but America is still the world's richest nation, and our expenditures on health are higher by one percent of our GNP than Holland's.[8] The explanation lies beyond heredity and money. Somewhere public and professional attitudes become very significant in making a difference in final health results.

Due to the Worker's and National Social Insurance programs the Dutch worker and his family are normally spared much of the burden of

[7] *New York Times*, December 31, 1972.
[8] *The Evening Bulletin*, Philadelphia, September 2, 1970.

their own economic security. After meeting the fixed, compulsory charges forced upon him, rather wisely by his government, the worker is free to spend liberally his take-home pay. National costs for these programs are high, about ten percent of the national income. Sickness and health benefits, and old age provisions take up the lion's share. Only half comes out of the worker's pocket, since the employer meets the remaining costs. Though the worker on average is taxed fully one-seventh of his income for social insurance premiums he is left almost totally free of any haunting insecurities. Most returns exceed his contributions. There is little necessity to save, though he continues out of the habit of thrift and government encouragement to do so. Fortunately there is in the Netherlands little of the insidious pattern of credit buying that entrap so many families in the American pattern of corrosive consumer debt. The Dutch social insurance system indirectly creates more equitable income among members of the society which can serve to sustain level demands upon the economy. Assured sustained incomes – often 80 percent of normal wages if unemployed, incapacitated, widowed or retired – the worker can maintain near normal buying levels. Such a prop to the economy can be provided by the retired who can remain near constant levels of consumption while not adding to possible excessive production.

There are growing difficulties in the foreseeable future for the Netherlands' social insurance programs. Increasing unemployment threatens the reserves of the Unemployment Fund. While efforts to keep down medical and health costs have been more successful than in the United States, inflation is still a problem. Most serious is the government's commitment to a future policy of greater coverage and benefits. It is proposing pension allowances by 1978 that will be equal to the prevailing minimum wage. Furthermore it proposes to raise the entry age from 15 to 18 and lowering retirement from 65 to 60. Thus payments will be put off while claims against the pension fund will be increased. Further taxing the Old Age Pension is the inescapable demographic fact that the Netherlands is becoming an aging society. Their past successes have made the Dutch optimistic, but only a continuously expanding economy can sustain such hopeful expansions of social benefits.

The Dutch social insurance programs are rather thorough in their coverage, generous in their application, efficient in their administration and practical in their funding. They reflect the national consensus that the government be responsible in attending to the social welfare of its citizens. America's experience with her doctrine of self-help has hardly been such to recommend it as the most effective means to deal with the social security, welfare and health problems of an industrial age. Prior to the New Deal social security measures, less than one-seventh of the population had adequate resources for security in their declining years. The living standards, nutritional and health care levels for a large part of her population are unworthy of the world's richest nation. Despite our attitudinal aversions we have been forced into wider social services. Without them there would be little to sustain the day to day economic demands, let alone the economic emergencies, so commonplace among our poor. Tragically we are still experimenting with piecemeal improvisations. Our strong individualistic values restrain us from adopting a philosophy of broad social concern that could then be implemented through intelligent legislation, effective administration and publicly sustained social costs. The Dutch have largely done this and have thereby virtually eliminated any serious need for a welfare program. The irony is that democratic-socialist nations of Northeast Europe are called Welfare States, when true welfare is most apt to be found in the more capitalistically structured America.

CHAPTER V

HOUSING AND URBAN PLANNING

The history of the building pattern of the Netherlands is still traceable in almost every old town or city. The Dutch built well and it is as if every house built was never torn down. Most buildings are of brick or stone; wooden frames do not fare well in the damp, inclement climate. Moreover, centuries of habitation and shipbuilding have denuded the Netherlands of any stands of hard timber that have existed.

The building pattern of most towns is a series of concentric rings representing different centuries of growth and architectural style. The center invariably has a square, serving as a market place, locale of the City Hall (Stadhuis) and providing setting and perspective for the dominating architectural pride of every community – the Kerk. Most, including the Domkerk of Utrecht, St. Bavo's in Haarlem and the Cathedral of Oudenbosch (a reproduction of St. Peter's) are tremendously impressive. The Cathedral of St. Jans in 's Hertogenbosch is a gem, among the most beautiful in the Netherlands. Its delicate columns and flying buttresses, its tracery, filigree and carvings represent a flight of fancy perhaps more gallic than Dutch. Churches of Holland are more apt to be solid, substantial and compact, for example, the Kerk at Steenbergen. All older communities have a centrally located religious edifice. Larger towns or cities have several. Few Kerken go back before the Golden Century. In the excesses of Calvinist purity many beautiful Catholic cathedrals were burned, destroyed or altered. One result is the absence of stained windows in the churches remaining. One felicitous exception is to be found in St. Jan's Church in Gouda. Next to the Kerk in tourist charm is the Town Hall. There are so many that attempting a list would be an injustice, yet the temptation to mention Gouda, Veere and Zierikzee is overwhelming.

It is not merely the Stadhuis but also the homes, that as contemporaries shared the square with it, that merit appreciation. They are invariably built out of brick with that characteristic stepped facade in lieu of the more conventional sloping eaves. Geometric patterns are often employed, enhancing their attractiveness. The windows are invariably gracefully elongated, the doorways carved works of art, the brass or wrought iron hardware a genre of their own – not to omit the protruding grillwork of the old oil lamps. They appear rather small or cramped for modern convenience. Perhaps the Dutch were smaller then for the dimensions of these earlier homes remind one of our own colonial dwellings. The cost of land and of building also determined the size and style of building. Many of these old homes were built on piling and reclaimed land. Consequently, even those of the rich, were narrow and squeezed together. Treacherous, steep "trappen" led the inhabitants to upper stories, and cantilever beams served for the entrance or removal of household effects. The absorbing charm of Dutch 17th century style is repeated again and again in the rows of homes fronting the canals in Amsterdam, Delft, Leiden, Utrecht and innumerable other Dutch cities. They convey an air of dignity and peace most restorative to the spirit. In the most busy and crowded of cities they offer asylum. Many towns, of former prominence, have been bypassed and remain almost as they were in earlier times. This is true of many former Zuyderzee communities – Enkhuizen, Hoorn, Blokzijl, Hindeloopen – now of minor importance after the sea's conversion to a freshwater lake. Similarly many of the towns of the Delta islands – Renesse, Brielle, Middelharnis, Goes – except for the summer homes of Victorian stamp, remain almost untouched. Very few cities are without their center core of ancient historical buildings. A few like Amersfoort, Kampen, Hulst were walled towns with the newer city mushrooming beyond the ramparts. Rotterdam's center, destroyed by the World War II bombing, is almost all 20th century with its renowned Lijnbaan – a non-traffic shopping area. Eindhoven, the home of Philips, has a center that is also quite commercial. Amsterdam remains delightfully quaint, with innumerable oases of the Golden Century throughout its later urban modernization.

The presence of these reminders of the past is important in the life and character of the Dutch. It gives them a sense of the past, a sense of continuity, a feeling of solid permanence that unconsciously becomes a part of Dutch life. The Dam, the Kerk of Delft, the Frans Hals Museum of Haarlem – these and others are Holland. There are unmeasured psychological dividends in a society where many members can make so close an identification with their past – and subtly, because of it, their present community. Throughout Holland these architectural reminders of the past give her inhabitants a sense of history and solid achievement. Coupled with the pride of workmanship, this perhaps makes up the most important ingredient of the national character – the feeling of being Dutch.

America by comparison can be accused of being without a past. In our feverish rush towards progress we have thoughtlessly destroyed or permitted the decay of most of the fine reminders of our colonial or Victorian past. Our inner cities have invariably degenerated into slums. There is some danger that this may also become true of areas in larger Dutch cities where older dwellings have become foreign worker homes. Still they are more apt to remain viable communities than their American counterparts. The bigger urban centers are also facing problems of inner city depopulation that have evoked actions on municipal, provincial and national levels. Middle class families are buying old houses in the quieter by-passed communities to escape some of the confinements, dissatisfactions, and impersonalization of 20th century urban living. Fantastic sums are spent for their restoration – providing homes of character for their owners while preserving them as examples of the way of life in the past. There are countless societies for the preservation and restoration of old dwellings and monuments of the past. There is even one to prevent the passing away of the windmill from the Dutch landscape. It is very impressive to see them stretched out like an orderly row of sentries at Kinderdijk. In America, we have the tendency to construct "shrines" – Mt. Vernon, Monticello, the Alamo – while to the Dutch their past is part of their daily lives.[1]

While numerous old city centers and by-passed towns remain with delightful and graceful architectural creations of Holland's Golden Age, most of the less enduring wooden shacks and poorly built brick hovels of the poor have long since succumbed to larger urban expansion. So too have most of the squat, ugly row houses built to provide

[1] In America we also have our "hallowed grounds" – battlefields of the Colonial, Civil and Indian wars. The Dutch have had too many battles during their long history to set these locales aside for solemn civic purposes. World War II was the first time since the Napoleonic era that Holland was disturbed by warfare but none of the battlefields have been preserved. They are, however, most scrupulous in maintaining the British, Canadian and American cemeteries that house the remains of their World War II liberators. It is an annual spring ritual for Dutch primary school children to place flowers upon the graves.

for the new urban poor of the early Industrial Revolution. These unfortunately were replaced by equally unimaginative row apartments. Many of these two to four storied "walk-ups" still exist, many without modern plumbing, heating and electrical improvements. They remain the prevalent unattractive, monotonous dwellings of the working class to be found near the center of the larger Dutch cities. Homes of the middle class built at this time are in contrast solid, substantial fortresses. They reflect the Victorian style of the late 19th and early 20th centuries. Located at what were then the city outskirts they are usually surrounded with small yards and gardens. Still the dominant style in most cities remain the two to four storied row homes whose monotony was not relieved until the advent of the Berlage style in the 1920's.

Berlage tried to solve the problem, which is still present, of obtaining maximum housing and an optimum of light, air and liveability in a limited space. Compared to modern Dutch dwellings, the Berlage creations now seem heavy. Yet his use of site arrangement, his angling of numerous units in a building complex showed imagination in the use of space, and permitted parks, trees and walks where they could never have existed before. He pointed the way, and with the development of modern building techniques a new concept of apartment dwelling was created. Some of the post-war dwellings are impressive in their creativity. With new techniques of foundation support, with use of new materials, particularly glass to replace heavier brick or concrete, modern buildings could go higher than ever. Initial post-war construction was timid and traditional. Though more modern in style and line, houses tend to be squat, less than four storied, in long rows and only half windows. It is the use of glass in modern Dutch dwelling that is so revolutionary. New buildings are taller, hardly any under six, while most in the larger cities are fourteen storied. They are invariably in park-like settings, with shrubbery, trees, walks and playgrounds for the children. Those in cities tend to be more constricted and more apt to be in rows.

A typical apartment arrangement consists of a kitchen-dining area, a living room and from two to three bedrooms – four of five rectangles within a rectangle. Light streams into every room for invariably one wall of each room is given over to an almost full window. Usually balconies lead off the front and the rear of the apartment. Dutch weather is most changeable and the sky represents one of the most exciting aspects of the Dutch scene. It is easy to understand the Dutch love of windows – and why they are always open to the outside world.

By far the gravest social problem in the small country of the Netherlands remains housing. Since 1945 1.8 million new housing units have gone up, a net increase of 80 percent over the two million homes in existence before World War II. Still the problem persists. The reason for the housing backlog is threefold – the Depression, the Occupation and the post-war population boom. Pre-Depression housing starts averaged 45,000 per year. During the Depression that dropped to 35,000 yearly average. With the Occupation housing ground to a halt. Moreover, bombing losses in World War II were high. It was not until 1948, with the impact of the Marshall Plan that construction revived, jumping from 9.000 to 35,000 housing starts in one year alone! During the 50's housing starts averaged approximately 50,000 annually, with close to 90,000 by the 60's. Present housing starts have reached the order of 150,000 per year.[2]

Heavy inroads have been made upon the nation's housing shortage, for while population during the post-war period increased by 25 percent, housing increased by over 35 percent.[3] The problem has been complicated by rising levels of housing expectations. As living standards rise the Dutch are expecting much more in the size and styles of their homes. At the turn of the century the average inhabitants per dwelling stood at 4.56. Today it is 3.57 and rapidly declining. The traditionally four room home is on the wane and five and even six room homes are more in demand. Only 15 percent of dwellings are four rooms or less. Fully 52 percent are five room dwellings, 26 percent being six-roomed. There is also a high level of owner occupancy – 70 percent fall in this category. Of course, like their American counterparts, such ownership is often shared with banks or mortgage corporations. While housing is now beyond the crisis stage severe shortages persist, particularly in the more densely populated west. Young couples are frequently forced to postpone marriage waiting upon apartment openings. Moreover, a considerable portion of the public live in what authorities consider substandard dwellings. Still rent controls have been removed from almost all the country but Utrecht and the two provinces of North and South Holland. This is a virtual miracle when one considers the housing

[2] 127.000 in 1970, 139.000 in 1971, and 156,000 in 1972. Approximately 75 percent of these in 1972 were one family dwellings. *Statistisch Zakboek*, 1973, The Hague, (Government Printers) 1974, p. 127.

[3] *Ibid.* Total units rose from 9.3 million in 1945 to 13.4 million in 1972, p. 16.

situation upon Liberation. Today one can say that a point of equilibrium has been attained – despite population rises that caused annual housing needs to rise from 55,000 at the beginning of the period to over 75,000 today.

Though some of the task was spurred by government building itself, the bulk was assumed by building associations and private enterprise. Immediately following the Liberation, the Central Government went into housing for a brief period of Reconstruction. It is the municipalities that assumed the role of government home builder. Today one-fourth of all buildings constructed is done by them. Another 28 percent of the building is done by the building associations. These are similar to our Building and Loan Associations. Originally authorized under the Housing Act of 1901, they collected money from workers to build less expensive multiple dwellings which they then rented out cheaply to them. From the beginning the building associations received subsidies from the Central Government, and they receive some form of subsidies to this day. These are largely in the form of guarantees on low interest loans, but often can be outright donations. Presently there is some tightness in the money market even for loans backed by government assurances. Building Association loans run at over an 8½ percent interest charge. Private enterprise accounts for the largest amount of construction, slightly under half of all building. This may or may not receive government subsidies. Usually preference is given to low cost housing construction. Average rent in government subsidized apartments runs in the neighborhood of $ 55 per month.[4] About 25,000 homes are built annually for the free market, without government subsidies or controls. These are usually more expensive homes for the wealthy or luxury apartments similar to condominiums. Still fully 84 percent of all housing built in the Netherlands today receives some form of government subsidy.

The control of building and the regulations of government financial help is carried out by the Ministry of Housing and Physical Planning. The Housing Act of 1901 has since been updated by the 1965 Housing Act. Under the terms of this legislation the government is required to present Parliament an annual report on Housing needs and progress.

[4] Interview, Dr. Schooten, Information Office, Ministry of Housing and Physical Planning. This figure needs upward revision to adjust to inflation and dollar devaluations. What is important is that ratios of earnings to rent levels probably remain somewhat constant.

The government is also empowered to set building codes and standards and to determine levels and allocations of subsidies. The power of the Municipal Governments to act in regard to housing was also expanded by this legislation. Municipalities have control over building by their right to grant or withhold building permits. Since they are required by law to submit planning arrangements to the Central Government, there are controls forcing building to conform to nation-wide plans. Municipalities must require building to conform to building codes that the Central Government has specified but has the right to add requirements of its own. Municipalities are responsible for funneling the government subsidies to those projects best meeting local needs and plans. They can also add subsidies of their own. Municipalities also have the power of condemnation – particularly of older dwellings or those not in keeping with city planning schemes. They are pressed by the Central Government to do so and where conditions make it feasible, the cleaning up and modernization of cities go on apace. The record is impressive.

Most important is that under the Housing Act, municipalities have the power of expropriation. This is fundamental to any hope for the implementation of planning. There is not quite the same degree of elaborate legal involvement that has become such a stumbling block to our own attempts at city modernization in America. There are legal safeguards for individual property rights but they are not insurmountable. There is more of a trend in the Netherlands towards concern for community needs than is found in America with its almost sacrosanct respect for narrow individual interests. As a result, hardly a city, town or community in the Netherlands has not undertaken substantial postwar redevelopment housing programs. Massive housing programs continue to this day.

The crush of population and the limitation of land makes a completely free market in real estate an impossibility in the Netherlands. From bitter experience the Dutch have learned, as perhaps we might, that uncontrolled free enterprise in such circumstances leads only to vicious land speculation and usurious rent practices. In order to combat exorbitant rents and substandard housing the Dutch were forced into regulation of housing and land use. The first applicable legislation was passed in 1851 as the Expropriation Act. It permitted local authorities to expropriate private property needed for roads, dykes, canals and other communal purposes. Though of little importance to housing or physical planning at that time, it became the parent law for ultimate government control in that it established the principle of public priority over unrestricted private property rights.

The key legislation in shifting emphasis from private property was the 1901 Housing Act. Radical powers were given to the municipalities in an effort to reform the scandalous housing abuses of the time.[5] There was even debate in the legislature to grant every worker an adequate home, but legislation stopped short of this. The law ended abuses of faulty construction, substandard dwellings, gouging rents, and it expanded low cost housing through financial aid to municipalities and housing associations. But most important was the inclusion of environmental standards in housing planning. Giving the authorities the ultimate power to condemn and expropriate was the sine qua non of true environmental control. Few, if any, understood the truly revolutionary concepts of the 1901 Housing Act when it was passed. In 1950 the Reconstruction Act was passed which gave wider powers of expropriation to municipalities than they enjoyed before. It also established heavier government standards on building and the loan of state monies for building. It virtually put all new building programs under the control of the national government. Without such legislation the low cost housing programs would be preempted in a private market by the more lucrative building to meet private demands. No building project without government approval could receive a building permit. Approval started with local authorities up to the Minister of Housing. Thus during the critical post-war years the most effective and equitable distribution of housing capacity was achieved. These restrictions are now being gradually reduced and the first relaxation of such controls began in 1965 when small private housing was permitted to again operate on a free open-market basis.

[5] The important provisions of this act are summarized below:

a. Each municipality must draw up local building regulations. Building permits are granted only when the building plan satisfies requirements.
b. Municipalities are authorized to combat poor housing by means of an order to the owner to make improvements or by powers of condemnation.
c. The process of expropriation was simplified.
d. Government financing of housing was exclusively to aid improvements of public housing.
e. Municipalities with 10,000 or more inhabitants must set up extension plans for use of land adjoining their built-up areas.

J. Bommer, *Housing and Planning Legislation in the Netherlands*, Rotterdam, (Bouwcentrum) 1967, p. 11 (with modification).

The Housing Act of 1962 updates that of 1901. Its important provisions centered on three areas – building codes, elimination of substandard housing, and the revision of financial aid programs. Building codes were made more uniform throughout the entire country. Moreover, wider powers of condemnation were given municipalities, now that substantial inroads were being eaten into the housing backlog. This facilitated a changeover to a mass-produced industrialized construction away from the older restrictive, expensive craft-oriented building methods. There has been extensive research in new materials, prefabrication and standardized construction methods. Private associations such as the Netherlands Standardization Institute, Economic Institute for the Building Industry, the Royal Institute of Engineers, the famous Bouwcentrum at Rotterdam, as well as the government sponsored TNO, have made significant advances in building modernization. The government has also undertaken programs to train construction workers in the new techniques and methods. The result has been a uniform modernized style that is immediately recognizable throughout the country.

Government subsidies were further liberalized. Not merely were municipalities able to receive government funds, but even Housing Associations, provided they were basically non-profitable in their operation. Owner-occupants of private housing were also eligible for loans or lump sum payments. Private builders and investors were given subsidies if they built low cost housing. Normally subsidies were in the form of low interest loans. There were other means to effect the subsidies. In the case of privately built low cost housing it could be a yearly contribution to the builder, or possibly a rent subsidy to the tenants. Under such arrangements it was the tenant, not the investor, who got the greater benefit of government largesse.

While the 1965 Housing Act merely extended earlier housing provisions, the Physical Planning Act was truly revolutionary in that it provided a means by which housing, style, siting, numbers, location would all become incorporated into a nationwide planning arrangement for better living. It had become apparent that by then population was so concentrated in the Netherlands that even effective municipal plans would not suffice to create a liveable environment for all her people. Particularly congested was the Randstad – the populated ring of Rotterdam, Amsterdam, The Hague, Utrecht, and surrounding suburbs – that had literally run out of space. One complication has been the in-migration to the Randstad which offers better employment opportunities than northern, eastern and southern Netherlands. This has offset much of the housing gains. In the Leyenburg, Mariahoeve, and Waldeck sections of the Hague over one-fourth of the new apartment dwellers came from outside the city. In the Alexanderstad and Ommoord section of Rotterdam less than one-third of the tenants are Rotterdammers. The building projects in the Randstad are ambitious, large complexes of multiple-storied (now twelve to fourteen high), glass and balconied structures, housing thousands in a single unit. Most of the complexes are attractive and convenient with elevators, central heating, maintenance services and easily accessible to shops, playgrounds, and schools. Only a deliberate nation-wide planning of high-rise inner city complexes or satellites complexes would permit this crowded land to remain attractive, pleasant, convenient or even liveable. Housing in North and South Holland shows that efforts toward this end have been rather effective.

Rentals in the Randstad prove to be rather expensive for housing is still a luxury in this crowded region. In the more fashionable Leyenburg, Mariahoeve and Waldeck sections of the Hague even the lower rentals take an inordinate share of these apartment dwellers' rather high income. These complexes are usually rented to older people, over 55, and working younger childless couples. Only a third of the tenancy are middle class families with children and non-working wives. The more expensive apartments have three to four bedrooms and a living room that could be 12 by 24 feet in size. The modest low rent apartment units in Leidschendam near The Hague still command one-fourth to one-third of occupant income. Moreover, as can be anticipated, the living styles are different from those of the higher rental areas. Three-fourths of those living in the lowest rental units have a family income of under $ 4,000 and are more apt to be young marrieds with children and a non-working wife. Even three-fourths of those in the more expensive apartments have incomes under $ 9,600.[6] Many feel that the rents they pay are too expensive. In a survey conducted by the *Haagsche Courant*,[7] only one-third of respondents whose rentals were in the lower range planned to stay in their present apartments. None of those in the higher range units said they would stay. While many complained

[6] Approximate dollar equivalent for *f* 10,000 and *f* 24,000 respectively.
[7] *Haagsche Courant*, June 1970.

of high rentals and poor service, barely 20 percent of the total *Haagsche Courant* interviewees was dissatisfied with the more modern and luxurious apartments. The pattern of tenancy in them is for higher income earners, older, including wealthier pensioners, or younger working marrieds. These last seem to prefer autos, appliances, clothes, more expensive vacations and other luxuries to families. This is clearly the trend in the more expensive housing. The more sophisticated apartment dweller has adjusted to modern apartment living and is far removed from the older traditional pattern of larger families, lower incomes and a more restricted standard of living. Over half of those interviewed from the more luxurious complexes said they would remain. Yet, many talk about moving to a private home, with yards and space, no close-by neighbors – and for a cheaper price – which is another aspect of Holland's (and possibly humanity's) penchant for holding on to impossible dreams.

Possibly Amsterdam is the best example of effective city planning in the Netherlands. It represented one of the most intriguing challenges to city planners. It is an old city, criss-crossed with canals, already densely built up with two to six storied row houses, and streets laid out long before the advent of the internal combustion engine. There was little the planner could do but to move out beyond the city limits. There they created minor miracles. People still move into the West of Holland for Amsterdam's charm and job opportunities lure thousands every year. Most move into the satellite bedroom communities that have been planned and built since the Liberation. The poorer tend to move into the densely crowded lower rental areas of the older city. But these are gradually being phased out. The demographic pattern of Amsterdam, like all cities, is for the native population to be restocked by outsiders who come to gain the material advantages of city life. Amsterdam, though declining slightly in population, is still obligated to provide housing far superior to the former pattern of crowding families into single and two room dwellings.

Amsterdam has had a seven hundred year old record of control and planning going back to the Golden Age. From its beginning the city was largely man-made, arising out of the sea – a city on sunken stilts. The princely merchant homes lining the planned concentric, semicircular Keizersgracht, Herengracht, Prinsengracht were all built on piling – as is much of the city beyond this. A plan drawn up in 1612 delineated the city's boundaries and provided for a ring of fortified walls. While Amsterdam prospered during the Golden Age, its history during the 18th and early 19th century was one of economic stagnation. It was not until 1866 that Amsterdam grew beyond the 17th century fortifications. The rivalry of France, England and the New World cut into her trade hegemony. The silting up of her harbors prompted the North Sea Canal project, but only the creation of the IJ canal made Amsterdam, once again, a viable port city. This and the industrial revolution led to a spate of unchecked city growth. With the passing of the 1901 Housing Act (which was largely directed at the housing abuses in that city) town planning began in earnest. In 1907 the Berlage Commission was created to plan for the southern extension of the city. The plan was not completed until 1917 as work was halted by the First World War. Still by the early 1920's the Berlage style prevailed. In 1928 the City Council created a Town Development Section in the Department of Public Works. Plans were also laid to set aside land for an extensive recreation and park area south of the city. A high water mark for city planning was reached in 1935 when the city adopted a General Extension Plan providing comprehensive planning for all of Amsterdam up to the year 2000. This was the first known full-scale scientific city planning project. Its provisions have already been fulfilled and with the new Bijlmermeer project the goals of the year 2000 have been far exceeded by 1970.

The General Extension Plan envisaged a population in Amsterdam of 900,000 by the turn of the century.[8] To provide total employment for the anticipated population, increase plans called for a vigorous extension of port facilities and the creation of a vast industrial park area in the southwest just beyond the city limits. Under the Plan's provisions the facilities of Schiphol were also enlarged making it capable of being one of Europe's busiest airports. There were also provisions for the enlargement of recreational facilities. The Amsterdamse Bos was created from the land set aside early in 1928. Plans also provided for construction of garden suburbs in the northwest and southern reaches of the city. In the northwest the communities of Geuzeveld, Osdorp, Overtoomse Veld, Slotermeer and Slotervaart have already been constructed and now house over 135,000. The largest single suburban

[8] Amsterdam's population beginning January, 1973 was 791,800. The population of the Amsterdam Urban Agglomeration was about 1,190,000. *Statistisch Zakboek*, p. 23.

complex, Buitenveldert, was built in the South to house over 35,000.

Buitenveldert is a fine example of effective planning. The entire complex was a single design. It was built at the southern tip of the city on land purchased in 1921. Leveling, grading and filling of the polder land began in 1955 and by 1959 the first buildings were open for sale or rental. By 1968 the project was completed but for some extensions that were not in the original scheme. The complex now has about 9,000 residential dwellings. Approximately 2,600 were built as subsidized dwellings under the Housing Act, 2,500 were sold outright, 2,900 were controlled rentals and approximately 600 were open market sales or condominiums. 31 percent of the dwellings was high rise, 56 percent three to four storied, and 13 percent was single villa type units. Land was too valuable for the suburban development characteristic of our own country. There are over 150 shops, usually clustered in convenient "Winkelcentrums." There is also a sprinkle of high rise office buildings employing over 2,000 workers.[9] Schools, hospitals, churches, shopping centers, motels, and parks are imaginatively interspersed to provide services and also to "break up" in part what could be an overwhelming mass of stone or concrete.

The design of Buitenveldert is far from the unimaginative square block row housing that was typical of Amsterdam's earlier construction. The entire area of Buitenveldert is ringed by open space. To the East are the river Amstel and a park; on the Northwest the open complex of Amsterdam's Free University. Three open avenues of parks – the Sportpark Buitenveldert, Gijsbrecht van Aemstel Park, and 't Kleine Loopveld – run the entire length of the complex from west to east. The central park avenue, Gijsbrecht van Aemstel, is from two to four city blocks wide, providing a delightful feeling of space right in the middle of Buitenveldert. The van Aemstel Park itself is bisected by a canal, is laced with walks, children's playgrounds, artistic creations and even a coffee shop or two. It is a pleasant and convenient place for relaxation of an afternoon or evening stroll. Cars are in evidence only at the limits of the park. Cutting the complex from north to south are three broad avenues, Buitenveldertse Laan, Rotterdamse Weg and Europa Boulevard. All three have walks and parklands. To the east of Europa Boulevard is another park area bordering on the river Amstel.

There are ample recreation facilities for the inhabitants of Buiten-

[9] Publication of "Bureau Voorlichting," Gemeente Amsterdam, undated.

veldert including playing fields with provisions for hockey, football, tennis and other team sports. Opportunities for rowing or sailing are found in the nearby Amstel, Amsterdamse Bos and the man-made Nieuwe Meer. Sprinkled throughout the complex are some 64 kinderspeelplaatsen or kiddy-parks. Buitenveldert is located close by the large Amsterdamse Bos which provides the inhabitants with open space, footpaths, horsetrails, sport fields, rowing lakes, sailing and all manner of recreational outlets, including private garden plots which are given out on yearly rental basis. Despite heavy population concentration, urban life can be pleasant and safe even for the young if planned for.

Traffic to and from, and through, the Buitenveldert complex, has been well planned. Since it was designed as a commuter bedroom community, provisions have been made for one parking space for each family. All the main roads going through the complex are four laned. The main east-west road – Van Nijenrode Weg – is four laned, designed as part of a future national highway leading to Utrecht. The Rotterdamse Weg is also a national highway. Planning is underway for a ring highway starting at the Coentunnel at Amsterdam west, circling in a loop close to Buitenveldert and coming west at the IJ once again in the direction of the latest Amsterdam housing complex, the Bijlmermeer. Adequate public transportation is also available from Buitenveldert into the city center. South of Buitenveldert lies Nieuwer Amstel and the housing complex of Amstelveen. These are also commuter suburbs of planned high rises that are beyond the city limits. These and the housing of Buitenveldert middle class families have the highest rentals found in the Amsterdam area.

The goals of the 1935 General Extension Plan, and those amendments to it made in 1939 were completed by 1968. Yet despite being thirty years ahead of schedule, housing and other service needs were still inadequate to the needs of Amsterdam's population. The influx of outsiders, earlier marriages, smaller families, and needs for commercial, industrial, traffic and recreational space demanded extensions to the original plan. City planners found they had to tie in their schemes with those for the entire surrounding area to fit into the guidelines laid down by national government under the control of the Rijksplanologische Dienst.

To meet housing needs the new plan called for residential expansion in Amsterdam North, in the south, the southwest, and replacement of older dwellings by more modern construction in center city. The suburbs north of the IJ – Nieuwendam, Oostzaan, Buikslotermeer – lie partially outside the city limits and will ultimately provide homes for 100,000 people. Those to the south also lie outside the city, and comprise largely an expansion of Amstelveen to provide for homes for more than 100,000 people. The largest single unit is Bijlmermeer in the southwest. It was originally planned to house 110,000 but even now as the construction is underway projections are being raised upward. There are also plans for an integrated satellite city of Amsterdam, not merely a bedroom community, to be built in the newly created south Flevopolder. It is difficult to project what housing needs will be met by modernizing the inner city. Much will be given over to office space and light industry, since employment increases in Amsterdam will be largely in these service areas. There are also plans for further expansion of secondary industries on the periphery of Amsterdam and the expansion of port facilities as well.

Since Amsterdam's ecology is fated to make her a center of control, administration exchange, fabrication and entrepôt, she has created a large commuter working class. This required thoughtful plans for the control of private and public traffic flows for the present and on into the future. Amsterdam is already a congested city. Fortunately more of its transportation is public than private – if one excluded the ubiquitous Dutch bicycle.[10] Most public transport in the city will continue to be the tram – the electric trolley which is clean, quiet, expeditious and convenient. There is not much point in enlarging the present tram network. Three radical lines start at the Centraal Station near the Dam. Each of these bifurcate at the outskirts of center city, carrying passengers to the bedroom suburbs Geuzenveld, Osdorp, Buitenveldert, Amstelveen, Astel, Diemen, Bijlmermeer. Most of these are over five miles from center city, still 80 percent of all destinations is within walking distance from stations and maximum travel time to outskirts from center city is 35 minutes.[11] Several years ago the need to build a subway was felt. Although opposed by several action groups, construction is already in an advanced stage.

[10] Presently bicycle traffic accounts for 60 per cent of inner city traffic. 25 per cent is public, largely the trams, and ten percent auto. This last is increasing by leaps and bounds. At peak load Amsterdam is almost a single traffic jam. What it will become with full use of the auto is too horrendous to contemplate.

[11] *Town Planning and Ground Exploitation in Amsterdam*, Bureau Voorlichting Gemeente Amsterdam, p. 13. May 1967.

Impelled by the growing increase of auto traffic, planners are forced to make provisions for access and throughway highways. Most of those already in existence, and those planned, are six laned, with off and on ramps instead of cross traffic. Most important is the planned ring motorway circling the entire city, connecting with the key national highways entering the city from the north, northeast, south and southeast. Ultimately the ring highway will tie in with the new highway going across the new polder land to Friesland and Groningen. There are also plans to create a ring road along the margin of the old city by enlarging upon the Centuurbaan and its continuation to the west. Already the IJ, the Coentunnel, and the bridge to Broek in Waterland are feeling traffic pressures. The IJ tunnel presently handles over 70,000 cars daily. Plans for additional crossings of the IJ are being drawn despite the exorbitant costs involved. The Dutch are usually energetic in solving physical problems, as the newly constructed Velsertunnel illustrates. Their experience with the sea and their soil has made their engineers both confident and imaginative. Subordinate higways are also planned for truck traffic. While elevated beyond the concentrated areas to avoid congestion they will go underground in the inner city. There are also plans for extensions of railroad lines, particularly towards Amsterdam southeast. New direct lines to the Hague and Rotterdam, avoiding the present ones going through Haarlem and Leiden are being planned.

In the new plans much emphasis is placed upon recreational needs of the growing population. As plans for each new residential complex are presented they are scrutinized to see if they contain the required recreational mix established by guidelines in part laid down by the national government and its new watchdog the Rijksplanologische Dienst. Amsterdam must be very grateful for the work of the foresighted earlier planners that gave them Vondelpark, Amsterdamse Bos, Beatrixpark and the lake area of Sloterplas. As Amsterdam grows towards the newly opened northeast, there is a possibility of excellent recreational development bordering on the future IJ lake. If the present discussions lead to a revision of the IJsselmeer scheme, and the Markerwaard polder is not created there will be even greater recreational resources for Amsterdam's inhabitants.

Modern Amsterdam's marvel of planning is the "fringe city" of Bijlmermeer. Like Buitenveldert it is a creation of a single plan. It is the culmination of the city's experience with planning and aims at creating a modern housing complex "Where the man of today can find the life style of tomorrow." Bijlmermeer is a compact area of approximately a mile and a half wide by two miles long, yet it will house in the neighborhood of 125,000 dwellers. The designers called for a totally integrated community supplying all living needs with the exception of industrial employment. Even some light industrial parks were later planned for along Bijlmermeer's southern rim.

Bijlmermeer lies just outside southeastern Amsterdam, surrounded by Ouderkerk, Abcoude, Weesp and Diemen. It was built on former polder land and hundreds of thousands of tons of sand were piped into the site, most of it from the IJsselmeer, though several lakes were created by taking sand fill from nearby areas. These conditions make construction costly – 50 percent of the cost of the project is bound up in preparation of the building site. Another factor burdening Amsterdam is the inflationary costs. Costs per square meter in the earlier northwest and southern developments were 37.5 guilders, while present per square meter costs at Bijlmermeer are more than 50 guilders. Even this is a bargain when one considers the project per square meter costs for leveling older construction and rebuilding in center city. These costs run at 600 guilders and more.[12] Mercifully the national government bears over 80 percent of the land costs and 50 percent of the costs of public work. Yet these prohibitive costs, with little or no profit, do not halt the building projects. Costs must be ignored for there are no alternatives. Moreover, under its communal housing programs, costs are spread throughout the population. Private capital cannot engineer nor pay the costs of such projects.

Lying some distance from the inner city, indeed separated from it by a narrow "green belt," considerable attention in planning was devoted to traffic accesses for the Bijlmermeer commuters. Four to six lane highways with non-intersecting crossovers, are under construction (some completed) to connect with the national highways and the ring motorway, also under construction. Bijlmermeer itself is ringed by a motorway, and the throughway from Amsterdam to the Gooi region runs as an elevated highway right through the project. Traffic from the expressways goes down ramps to smaller and less trafficked roads bisecting the project. These are few in number, several blocks apart. Cars are parked in specially built multiple storied garages that match surround-

[12] *Town Planning and Ground Exploitation in Amsterdam*, Bureau Voorlichting, Gemeente Amsterdam, p. 37, undated.

ing architecture and are located within walking distance of the commuters' flats. In addition a highspeed railroad line is under construction running from Amsterdam to Utrecht with convenient stations along the western rim of Bijlmermeer. A similar line runs from Amsterdam to Diemen and beyond, with an equal number of stations on Bijlmermeer's eastern rim. Commuting time from Bijlmermeer to Amsterdam Center is designed to average thirty minutes.

One of the most pleasing aspects of Bijlmermeer is the illusion of space it yields to the observer. Ironically one is more impressed by the horizontal flow of lines rather than height. Possibly this is the effect of the running balconies and the unusual length of the buildings. Only as one comes face to face with an apartment unit is one aware of its surprising height.

With this feeling of space and without auto traffic the centers of the complexes should be pleasant and restful. Aside from the few smaller streets and the central highway overpass planned and under construction, there are nothing but parks, recreational areas, shopping centers and walks in center Bijlmermeer. There is also to be a centrally located high school complex, a consolidated school offering the best of facilities for Bijlmermeer's children. These schools will be added to, if necessary, by the separate confessional schools that are so intrinsically a part of Dutch education. Cycle, walking and bridle paths will form a network of "indoor streets" throughout, and children will be able to roam about without the menace or noise of the automobile. No one will be more than one-quarter of a mile from all facilities. Sheltered lanes will permit walking during rainy weather, a not uncommon phenomenon of the Netherlands.

There are some depressing aspects in seeing so many people living so close to one another. Bijlmermeer outwardly seems to be a practical and pleasant adjustment to the dilemma of population density and space limitations. However, many of its inhabitants tend to form a differing opinion in the long run. Certainly it represents efforts which attempt to cope with the gross dislocations, ugliness and humiliations that arise from unplanned urban growth. The lifestyle choices of man are rapidly dwindling and Bijlmermeer promoters may have said more than they realized when they spoke of the man of today living in the world of tomorrow.

While Bijlmermeer is the most ambitious housing complex in the Netherlands it is far from being unique. We may boast of our Reston and Columbia villages housing some 20,000 in newly erected suburbs, yet every Dutch city and innumerable towns have more compact housing built as entire communities outside their walls. Alexanderstad in Rotterdam, Mariahoeve in The Hague, Osdorp and Slotermeer in Amsterdam, are only the more familiar. Zoetermeer, eight miles east of the Hague, was built as a concentrated suburb complex to house in part that city's near 100,000 population overflow. It is conveniently located on the Rijksweg, or superhighway, leading to Utrecht. It is just being finished with housing facilities for over 30,000, with its own industrial park providing employment opportunities for some of the new inhabitants. The Dutch pattern is not to go into suburban sprawl but rather to create compact, convenient housing with acceptable rent levels. This is an understandable goal despite the higher building costs that it entails. Suburbia, the mania of America that is largely responsible for the decay of our cities, is a luxury the Dutch can ill afford.

Undoubtedly the shortage of space induced the Dutch to prize it all the more. While we would "eat up" our land, the Dutch would conserve theirs. Nor is their culture to be overlooked in their scheme of things. In their housing and urban planning, historic edifices, beauty of design, and healthful environmental supports take high priority. The American emphasis upon material comforts tends to override these considerations. We so often measure our blessings by our material possessions and not by the peace, quiet, beauty and sanity of our lives. This is so often reflected in our cities. Perhaps we are beginning to appreciate the impact of these intangibles upon our happiness.

Unfortunately the blight of our damaged cities is already with us and we can restore their history, beauty and liveability only at great costs. We are unprepared to submit private property rights to the public weal to the degree which the Dutch are. We may confiscate property for highways through our cities, for the auto seems sacrosanct, but rarely to build public housing or parks. We have yet to learn that private ownership of individual homes in suburbia may not be the wisest solution to our nation's housing needs. Nor have our government sponsored housing and urban renewal programs been effective enough to merit public confidence or support. That such problems in the Netherlands are far from being solved is obvious, but one cannot say that Dutch accomplishments have been as disappointing as our own. The rebuilding of her cities into habitations of pride is not America's foremost achievement.

CHAPTER VI

PHYSICAL PLANNING

The ecology of an environment is a function of population density and the use to which the natural resources of the environment are put. In the densely populated, over-used environment of the Netherlands, pollution and congestion problems become inescapable. Despite our human propensity to wish it away there is a growing neo-malthusian inevitability in the world's demography. The only possible escape is the technological – the mastery of the environment by altering it from a natural to a totally controlled and man-made living space. The Dutch have not been reluctant nor lacked the confidence to move towards this direction. Much of their environment is man-made and much of their lives is controlled. Perhaps no other small region has so intense a technological development. Dutch experiences may be profitably reviewed by others whose future may take a similar path.

That the Dutch are well along this path can be seen by their creation of a National Physical Planning Service (Rijksplanologische Dienst) to integrate, to advise, and to control the development of their entire country. Planning in the Netherlands is no longer merely local, municipal, regional – but nation-wide and near total. At least such is the intent of the Physical Planning Act of 1967. Prompted by the growing complexities of modern technology and her demographic pressures – particularly in the Randstad region – the Dutch were forced to expand planning and control services. Her economic situation led her early towards planning and control of trade, industry and agriculture. This helped promote a secure economy that expanded population; creating problems of housing, pollution and traffic, recreation and living space itself. At this point the Dutch were virtually forced into near total planning.

Population density in the Netherlands is over 900 per square mile – the highest in Europe. One-half of her population lives in one-fifth of her land. The Randstad includes Rotterdam and Amsterdam, each with populations near one million; and The Hague and Utrecht with half a million apiece. Despite her vigorous plans for land expansion and her intensive use of former wasteland there is no possibility of avoiding greater population density. As it is 70 percent of Holland's total land area is given over to cultivation, one of the world's highest ratios! From 1900 to 1950, 150,000 acres of land were wrested from the sea and 1,130,000 acres from wastelands. During the same period 400,000 acres were commandeered for roads, canals, airports, industry, and housing. Because of this and her population increases, there was a net loss per person despite the increase of available total land. In the period mentioned agricultural land declined from 1.13 to 0.6 acres per person. Only another 565,000 acres from the new polderland can be hoped for in the coming years. High agricultural yields are possible only with intensive scientific farming which entails a growing risk of diminishing returns. Moreover, mechanization and other agricultural efficiencies call for consolidation creating a smaller farm labor force. Those displaced need to be absorbed in industry or services. Presently about 18 percent of the population is engaged in agriculture with anticipated levels for 1980 being under five percent.

Since 1950 the Ministry of Economic Affairs has been working to enhance the efficiency of industrial productivity in the Randstand areas and in creating new industries in the "development areas" of the Netherlands' northeast and southwest. Pivotal to Holland's economic success has been her geographic position vis-à-vis the population concentration centers of Europe. The Amsterdam area with IJmond and Velsen; Rotterdam with Pernis and Botlek, (and now Europoort) are outstanding examples of Netherlands' creation of a totally technological environmental complex to capitalize upon her geographic centrality to west and central European markets.

Her agricultural and industrial successes promoted a large service sector in the economy. Today 45 percent of the work force is engaged in trade, banking, insurance, education, research, medicine, government work, etc. The standard of living is high but the environmental costs are equally high. Housing, recreation and the preservation of nature are important problems. Pollution and traffic problems are increasing. With space at a premium there is intense competition for its use. It is this that has called the Physical Planning Service into being – to balance out the conflicting claims to land and to oversee its most efficient and beneficial use.

The concept of national planning did not spring up overnight. The Housing Act of 1901 implied some concepts of national standards in use of building space. By 1931 legislation was passed to facilitate arrangements for regional planning. During the Occupation the government proposed, prematurely, some national planning. There was neither enough data nor experience to project a comprehensive plan so early. Under the prompting of government leaders and private associations a temporary act in 1950 divided the Ministry of Housing into a Housing and a Physical Planning branch. The Physical Planning Act of 1957 created a formal Physical Planning Service agency. It was to coordinate the broad outlines of National Physical Planning policy. Policy was to be determined by a Physical Planning Council. Most members were professionals and so important was this council in the eyes of the government that no less a figure than the Prime Minister served as its chairman.

Public support came from private associations, the Netherlands Institute for Physical Planning and Housing being the best example. It has close to 1,000 members, mostly professionals and students. The Netherlands Institute was founded in 1918 in response to the housing problems of that time. It was instrumental in effecting the regional planning legislation of 1931 and 1941. It played a role in town planning for the Zuiderzee towns. Its membership increased greatly following the government's creation of a Physical Planning Service. It assists the government in its efforts for planning by disseminating information and pleading for public support.[1] Inevitably other groups – associations of contractors, engineers, industries, the Netherlands National Auto Association – all with an interest in how physical planning would affect their activities, have found a way to make an impact upon the Physical Planning Council's decisions.

Following procedures established as early as the 1901 Housing Act, local municipalities are required to devise their own planning schemes.

[1] The institute has its own research and planning departments, an information center, study groups on the sociological and legal aspects of physical planning, and conducts nationwide and international seminars. It conducts liaisons witho ther international planning groups such as the English Town Planning Institute, the Deutsche Verband für Wohnungswesen, Städtebau und Raumplanung, and the American Institute of Planners. It publishes its own house journal, "Stedenbouw en Volkshuisvesting." Its publications often prove to be quite effective such as its 1965 *Physical Planning in a Motorized Age* which prompted the government to emphasize traffic in its *Second Memorandum of Physical Planning Service* in 1966.

While larger cities generally have their own staffs, smaller municipalities employ the planning services of private firms. Of the Netherlands' 900 municipalities, 70 percent already have approved master plans, 25 percent are in the process of planning, with 5 percent without such plans. These last are small rural communities with few if any environmental problems. The Provinces review and regulate the plans submitted to them. There are also regional planning bodies, the IJmond and Rijnmond authorities and the Waterschappen for example, that also plan environmentally in cooperation with the Provincial and National Planning services.

The Government Physical Planning Service (Rijksplanologische Dienst) serves as a counsel and watchdog for overall national planning goals. Final decisions are left to the Crown. There was until recently little of the political in these decisions and the future pointed to the continued increase of professional objectivity in planning. As in other newly created Dutch agencies, professionalism was the watch word. Mr. Quene, the first director appointed to head up the new Planning Bureau was a university professor. At the present early stage of national physical planning there are, as can be expected, some elements of confusion, hesitation, and uncertainty. There was no immediate aim to consolidate all the previous separate planning arrangements, such as the IJsselmeer, or Delta projects, into one over-all national scheme. Moreover the Physical Planning Service acts more as an advisory than a command body. However, the logic of circumstance will predictably move the Physical Planning Service to a role of greater prominence in the Netherlands' future. It will surely be a career avenue for her bright young men. Moreover as the present levels of public concern and criticism intensify, a considerably greater democratic participation in the physical planning process can be forecast.

An early report of the Rijksplanologische Dienst was its 1960 "Memorandum Concerning Physical planning in the Netherlands." This report centered on the problem of "Randstad Holland" showing the bureau's broad-based concern for important areas of the Netherlands' future ecology. It urged the preservation of the historic city centers of this region. While recommending intensive improvements within already built-up areas it would maintain as much as possible the present "green" area within the Randstad's center. To aid in this goal it recommended the centrifugal expansion of Randstad in newly created living spaces of IJsselmeer and the Deltaland.

By 1966 the Rijksplanologische Dienst had ready its "Second Memorandum on Physcial Planning in the Netherlands." This report accentuated the effects of accelerated rates of technological complexity upon environment and society. Perhaps the most important of these to the professional planners was that the Netherlands had now reached a point where almost every planning decision made sets an irrevocable train of events. This implied an understanding that options close, once a road is taken. How well it is understood by the government or the citizenry is another question.

The 1966 report based on Central Bureau of Statistics figures projected a population of twenty million by the year 2000. It foresaw a further decline in the numbers engaged in agriculture and the greater movement towards urban centers. This meant a further demand upon valuable space for industrial parks, housing, traffic needs. The report showed considerable maturity in emphasizing that the physical problems were far outweighed by social uncertainties of tomorrow's Netherlands. The concentration of tremendous numbers of people in newly constructed work and living environments posed the problem of the creation of new communities. What would happen to the Dutch society previously so sound and stable in its earlier self-contained feeling of community?

Unable to answer this question the report turned itself to the physical aspects of future growth. Its solution to increasing urbanization was its program of "concentrated deconcentration." This was the careful spreading out of dwellings in a concentrated use of space. Commuter suburbs serviced by efficient public transport would be built outside city limits with their own nuclei and high rise units. There would be "green belts" isolating the new complexes which would be a wholesome mix of single family home with multiple dwellings. It urged moving population out of the high density Randstad to the government encouraged growth areas of the North, Overijssel, Central Limburg, North Brabant and southwest Zeeland. The government target by 2000 for the Northeast was a population of three million, and three-fourths of a million for Southwest Zeeland. To encourage migration the government began decentralizing its own offices in The Hague. While not ignoring measures to develop the economic growth of the country as a whole, it particularly emphasized improving the physical plan of her entry port areas of Rijnmond, IJmond, Eems and the Schelde.

The 1966 Memorandum contained a special section devoted to auto

THE NETHERLANDS NATIONAL PHYSICAL PLANNING BUREAU

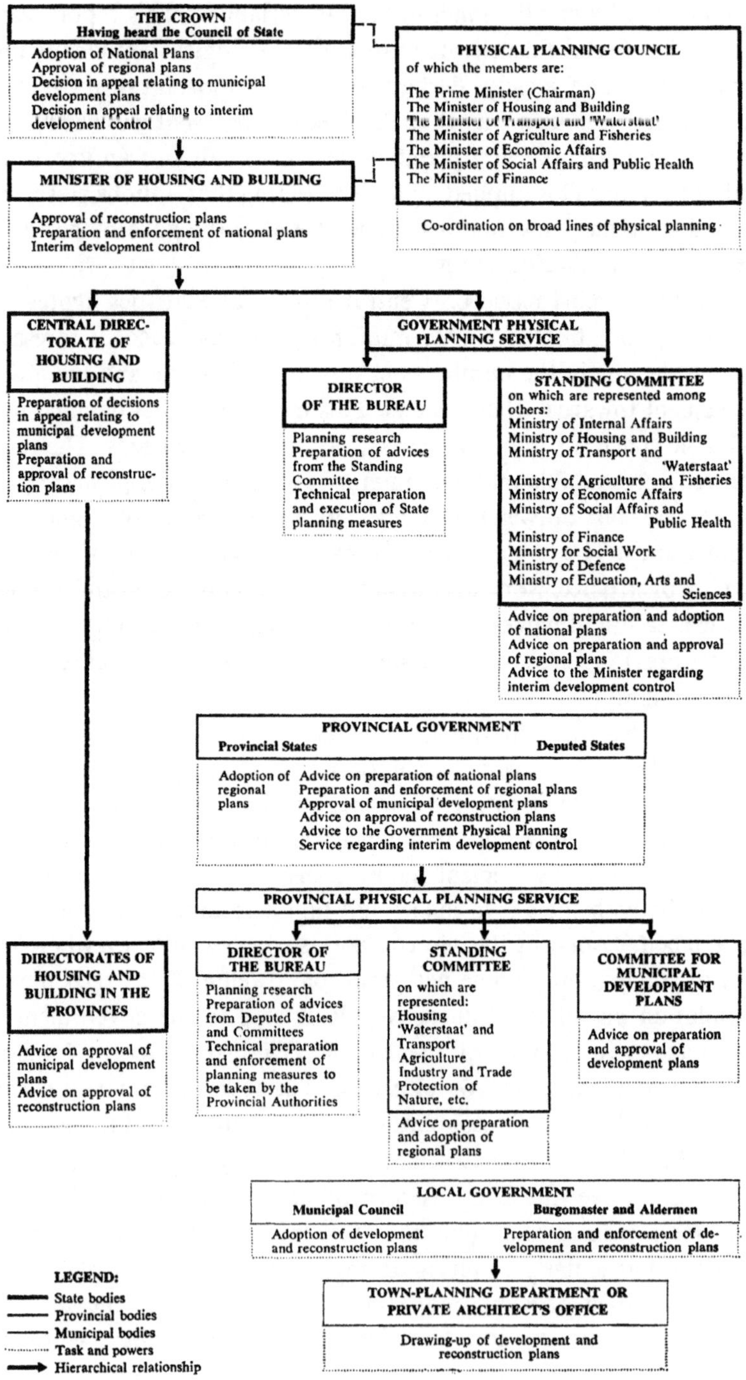

traffic which it considered as one of the nation's most difficult problems. Its predictions were that there would be from six to seven million autos in the Netherlands by 2000 – a ratio of one to three in motor density. It projected threefold expansion of the nation's road network to an ultimate 3500 mile total. It would create a system of clear-cut functional differentiation – motorways (Autowegen) as main distribution roads, secondary distribution roads, and residential streets. Significant was the Report's insistence that the government maintain and expand public transport. The planners were convinced that in the long run the cities would be unable to cope with mass motorization. Some talk of forbidding the use of the auto in city centers was suggested.

Another section was devoted to measures to protect the natural environment, the guarding of water purity, the elimination of waste, the pollution of air, and the elimination of excessive noise. In this the Dutch find themselves in the same dilemma of all within industrialized society. The factories that create the smog, the jets that bring sonic booms, are also the avenues of wealth to the society. The Memorandum revealed Dutch determination to do what could be done. Strenuous recommendations to preserve the natural beauty of unspoiled areas were offered. As elsewhere in industrial societies, natural areas are under heavy attack. The Veluwe, Netherlands' largest single open area, was threatened by water authorities wanting space for a reservoir, road builders proposing a national highway through it, and the military that would preempt it for their maneuvers.

The 1966 Report was more a suggestion than a plan, a "go slow" approach similar to so many areas of decision in the Netherlands. The law had given much of the responsibility for earlier planning to the municipalities and provinces, as well as subordinate government units, and the Rijksplanologische Dienst, as a newcomer on the scene, was careful not to upset previous institutional arrangements. The Minister of Housing and Physical Planning when representing the 1966 Memorandum to Parliament took pains to remind the legislators that it was "... not a 'National Plan' and it is impossible to read off the future from it. What it gives is a development pattern which the government sees as the most desirable pattern on which to base its policy."[2] Hardly total, hardly prescriptive, it still seems moving in the direction that ecology demands for this late phase of the 20th century.

[2] *Summary of Second Memorandum on Physical Planning*, Information Service, Ministry of Housing and Physical Planning, The Hague, 1968, p. 8.

The National Physical Planning Agency's 1968 yearly Report[3] reveals the extent to which concern for planning had grown. The Agency moved deeply into areas of environmental protection, industrial expansion, agricultural alteration, housing, traffic and pollution control. It began its report by incorporating the progress of the Delta, IJsselmeer, Lauwerszee and Waddenzee projects and their impact upon total environmental planning. Earlier these would have been the province of individual authorities. Surprisingly the Report took no exception to the proposal for a low level road dam to the offshore island of Ameland, nor the plan to eventually dam off the Waddenzee itself. Before taking its final position on the Waddenzee project the Physical Planning Agency coordinated the reports of the Rijkswaterstaat, the Wadden Committee of the Royal Netherlands Academy of Science, and the Wadden Area Working Party. The government decision included local economic interests such as the fishing industry, nature groups, and local inhabitants. This region is one of the few untouched areas of the Netherlands – a sanctuary permitting man to be one with nature. Public opposition to the proposal runs high and may ultimately defeat it.

Though plans for industrial growth were seen as necessary, efforts were made to keep the unwelcomed effects to a minimum. The proposal of the Minister of Housing and Physical Planning to lay out the new industrial areas of Europoort and Moerdijk as far from residential areas as possible was endorsed by the Report. The Report proposed further expansion of other industrial-seaport areas. The Rijnpoort was to be a new complex on the north bank of Rotterdam's waterway. Another would be at Terneuzen now made important by the new canal to Gent able to handle ships of 50,000 tonnage. The Gent-Terneuzen canal, planned, engineered and built by the Dutch, was largely financed by Belgium. This country has not usually been as energetic as the Netherlands in re-capitalizing its industrial infra-structure, and Gent was slowly declining, as its water access became more outmoded.

The Report discussed the plight of Dutch agriculture. Throughout Europe farming had become a "sick" segment of the economy. The Report accepted the findings of the "Mansholt Plan" of agricultural reform written by the Common Market's European Agricultural Commission in 1960. The Commission had made it quite clear, even that

[3] *Annual Report, 1968*, National Physical Planning Agency, The Hague, 1969.

early, that neither price nor market manipulation, but only structural reforms could solve Europe's agricultural difficulties. There were just too many farmers, with too little acreage, too antiquated equipment, and too uneconomical marketing arrangements to enable them to compete effectively with the larger, more efficient combines. The abandonment of marginal farms would displace over five million farmers in the Common Market. The Netherlands was already well along in this change but was still feeling the consequences of further modernization among her competitors. The Report urged further curtailment in areas of Dutch agriculture and called upon the government to assume the task of fitting the displaced farm workers into industry or services, as well as to help them adjust to the life-styles accompanying this shift.

The Report also gave information on the housing situation in the Netherlands, which fortunately was approaching adequate levels. To the planners, this meant that housing authorities could now turn their attention to the elimination of sub-standard dwellings. The Ministry of Housing and Physical Planning placed about one-third of a million present homes in that category and called for the replacement of 35,000 such homes annually. It also suggested that the government provide funds to improve a quarter of a million homes which would then be more habitable. Housing still remains a problem.

The 1968 Report reveals that the government spent considerable energy wrestling with the problems of traffic and transportation. A State Road Plan, confirmed by Royal Decree, calls for doubling of existing roads. A planned additional 1850 kilometers would be mainly highways to extend the present 750 kilometers of highways. Environmental pollution also received much government attention. An air pollution bill was submitted to Parliament that year which provided for automated monitoring devices to assess and control air pollution at its source. It provided for fines and penalties to offenders. The bill received considerable public support. For some reason the Dutch seem inordinately concerned with the problem of fresh air. Another nuisance acted upon this year was that of excessive noise, when a special department for its control was established under the Government Aviation Agency. Airport officials at Schiphol, evidently more sensitive to public feeling than their American counterparts seem to be, established a Noise Information Center to help curb this new bane of technological civilization. One rather creative innovation was a law providing for regional "car dumps" to contain the spread of this eyesore.

Water purity was a priority problem. The Report accepted the Waterstaat's *Memorandum on Water Management in the Netherlands*, advocating the creation of two major water preservation reservoir areas, one in the IJsselmeer, the other the Zeeuwsche Meer. The Parliament also passed a water pollution control law. The Report deplored the polluted condition of the Rhine, much of which was due to the French, who, despite having signed the "European Water Charter" in Strasbourg that year, continued their non-cooperative practices of industrial discharges into the Rhine. Particularly vexsome was the continued clandestine emptying of residual oil by river barge traffic. The government recommended tighter safety controls of vessels in port areas and invited international cooperation in attempting to control and assess legal responsibilities for carriers. There was little international response. It had more success in domestic legislation requiring the purification or selection of oil to reduce sulfur content and other harmful or noisome pollutants.

The National Physical Planning Service Report made vigorous protestations against the seemingly inevitable industrial and urban encroachments upon the nation's "green" zones. Most inroads occurred in the proposed "green belt" buffer nearest to the already developed Randstad – notably in the Midden-Delfland and Spaarnwoude regions. If population, industry and urbanization continue to grow it will be impossible to keep these areas intact. The problem has already cut across former municipal or provincial boundaries and jurisdictions. 1968 saw the formation of wider regional administrative authorities with more inclusive functions and powers. The role of larger cities have also been expanded, encompassing regional tasks of housing, industrial traffic, welfare and cultural responsibilities.

From the implications of the National Physical Planning Report and the Government's response to it, it would seem as if the entire Netherlands is to be a well planned, hopefully pleasant, functional city. What is disturbing is that the Report, and possibly government policy as well, does not think far enough ahead into the implications of the ecological conditions they describe and the technological solutions they propose. Too often ignored was the issue of loss of community sense and belongedness that continuously expanding artificial environments would create.[4] Moreover how much of a burden can the physical geo-

[4] The Report did use other studies to show that a sense of belongedness and

graphy sustain? Can one continue to alter the natural environment indefinitely?

Subsequent annual reports of the government reveal an inclination to increase the role of physical planning in Netherlands affairs. As population, industrialization and attendant problems increase, there is greater awareness of the importance of total planning to mediate conflicts between the need for industrial development and what the Dutch call "collective welfare elements" – recreation, nature areas, and attractive living space. Government awareness was accompanied, and often prompted by growing public interest and criticisms of national planning efforts. Planning in the Netherlands is no longer solely an affair for town planners and university specialists. It is a part of the Dutch way of life.

It is not inappropriate to continue a description of planning in the Netherlands with the most popularly known aspect of Dutch life – Holland's victory over the sea. Her struggle against this enemy has a long history. Her achievements are outstanding, and continue to this day. Success came from planning and cooperation. Planning for the acquisition of new land, building of new dykes, fortification of the old, dredging of the canals, control of water levels, erection of bridges over canals and roads upon dykes, and the elimination of wastes and sewage through the waterways – all this is familiar routine in the Netherlands. These tasks are under the administration of the Waterstaat, possibly Holland's oldest agency for planning and control. There is no precise translation for the word Waterstaat. It covers any activity related to control of water. The Waterstaat is a nationwide authority with great powers. In emergencies it has priority over other branches of government and can commandeer labor, private property and all manner of public resources.

Such an authority is a logical outgrowth of Dutch geography. A tidal land, one-half of her 13,000 square miles below sea level, Holland must place priority upon her security against the sea. Complicating the geographic problem is the fact that three major rivers – the Rhine, the Maas, and the Schelde debouch in these lowlands. Water must be so controlled that at high tides floods occur neither from the sea or the swollen rivers. At low tide the rivers must be emptied into the sea.

citizenship responsibility does deteriorate when the environment becomes too large or impersonal. One study set 35,000 inhabitants as the optimum point for a viable urban community. There seems to be altogether too little research in this aspect of Dutch life.

There are over 2000 miles of dykes and dunes, with untold locks and pumps, protecting the Netherlands. Some 1100 miles of these face sea and lake; the remainder rivers, most of which flow quietly well above the surrounding landscape, trapped within their dykes.

Since 1200 Holland has increased her land area by 2500 square miles mostly below sea level. For all their outward calm the Dutch are living dangerously. Battles have often been lost against the sea as periodic flood disasters occur. Ironically there has been no significant increase in land before the 20th century, for Holland has lost through erosion and floods an area equal to what she has gained. Her situation is still such that she must forge ahead in order to keep her gains.

Important progress in land reclamation began when the technical progress of the 16th century – the revolving windmill, water pumps, etc. – made large- scale drainage operation possible. Fortunes from trade and colonial empires seeking local investments outlets also came at this time. The result was a rash of land reclamation projects. The creation of the polders Beemster (1612), Purmer (1622), Wormer (1626) and Schermer (1635) developed the techniques of land reclamation and established the profitability of such operations. Two centuries later in 1854 the draining of Haarlem Lake was the largest project of its kind and created 45,000 acres of excellent farming land. It also spared Amsterdam from possible commercial extinction. Poor drainage of the lake was creating a gradual inundation of that city. A famous naval battle occurred on the former lake where Schiphol is now located. The airport is fourteen feet below sea level and thus one encounters the astounding sight of ships in the canal gliding above the resting jets.

The significant gains in land reclamation belong to the 20th century and are still in progress. The two most important are the Zuiderzee (IJsselmeer) and the Delta projects. The Dutch were simultaneously working on a smaller reclamation project in the Lauwerszee. They are also carrying on extensive innovation of port facilities at IJmuiden, Rijnmond and on the Schelde and Delfzijl. These are controlled by separate authorities though there is much cooperation and liaison with the Waterstaat.

The Zuiderzee project was the child of Dr. Lely who presented his plans to the Dutch government as early as 1918. The food shortage arising in Holland during World War I and the disastrous floods of 1916 made the government aware of the need for some action. It was the Depression years that helped bring the project into birth. It was

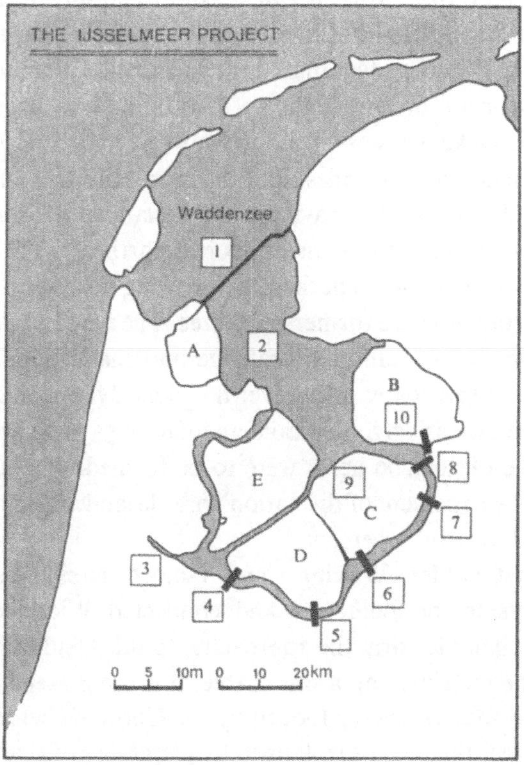

A. Wieringermeer	(46,000 acres-approx.)	
B. Noordoost	(105,000 acres-approx.)	
C. Oostelijk Flevoland	(120,000 acres-approx.)	
D. Zuidelijk Flevoland	(94,000 acres-approx.)	
E. Markerwaard	(132,000 acres-approx.)	Under Reconsideration

1. Afsluitdijk	6. Harderwijk Bridge
2. IJsselmeer	7. Elburg Bridge
3. Amsterdam	8. Kampen Bridge
4. Muiderberg Bridge	9. Lelystad
5. Nijkerk Bridge	10. Emmeloord

seen by the government as a partial solution to the poverty and unemployment of the time.

In January 1927 work began on the Enclosure Dyke (Afsluitdijk) that would convert the Zuiderzee into Lake IJsselmeer. The dyke, 20 miles long, stands 25 feet high, is 300 feet wide at sea level, and serves as a bed for a four lane highway. A rail line was also projected. The Enclosure Dyke was completed in May 1932. As a result the shoreline was shortened by 200 miles, and the more direct connection between North Holland and the provinces of Friesland and Groningen augured well for the future of the less developed northeast. The creation of the fresh water IJsselmeer checked the growing salination of agricultural areas surrounding the former Zuiderzee. The new Lake's freedom from tidal movements facilitated drainage and made impolderization less difficult and less costly. Moreover the older dykes could now be maintained less expensively. New polders totaling 555,000 acres and a fresh water lake of 300,000 acres were to be formed. The new land would account for 10 percent of the nation's arable land and increase Holland's total land area by 7 percent.

The first polder, Wieringermeer (50,000 acres) was created from water outside the Zuiderzee and connected Wieringen Island with North Holland to form the approaches to the Afsluitdijk. Seventy percent of the reclaimed area was arable. It is the present site of over 500 farms and four villages. Flooded by the Germans before their evacuation in April 1945 it was reclaimed by December of that year and crops were obtained from the recovered land in 1946. This flooding revealed another advantage of the Zuiderzee project. The presence of the Afsluitdijk meant that fresh, not salt water, flooded the land. Had it been otherwise the reconstruction would have taken years.

The second reclamation project completed was the Noordoostpolder (12,000 acres). Begun in 1937 it was completed in 1942 during the occupation. This polder was also willfully flooded by the Germans. It is 80 percent arable, has over 1600 farms, ten villages and a central town, Emmeloord, with a population of about 30,000. Today it already has the impression of an established land and society.

The third polder, Oostelijk Flevoland (135,000 acres) was begun in 1953 and completed in 1957. It is 97 percent arable and one of the most fertile regions of Holland. The original plans were to use the polder almost entirely for agriculture, but population pressures in the surrounding "old" land altered such intentions. Instead of the original plan of ten small villages with a major town on the design of the Noordoostpolder, there are just three larger urban areas, Dronten, Swifterbant, Biddenhuizen, and a large city, Lelystad, with a planned 30,000 to a possible 100,000 population. Oostelijk Flevoland was also developed with a more improved plan for recreation than was the Wieringermeer or Noordoostpolder. Almost the entire eastern coast of the polder bordering Drentenmeer and Veluwemeer has been made into a conservation and recreation area – parks, beaches, marinas – to serve the populated areas around Zwolle, Apeldoorn, Amersfoort and Utrecht.

Two polders remain, Zuidelijk Flevoland (108,000 acres) and the largest of all, the planned Markerwaard (150,000 acres). Zuidelijk Flevoland began in 1959 and was enclosed by 1969. Sowing of grass crops earlier to prepare the soil for agriculture was done from the air the following year, while the land was still waterlogged. Soon a bridge at Muiderberg was built that directly connected Amsterdam traffic with the Northeast of the Netherlands through the polders. Plans for Zuidelijk Flevoland are in flux attendant upon new population changes. It is the feeling of the planners that the urban sprawl from Amsterdam will probably consume much of this polder land. A fourth of it will be given over to urbanization, another fourth for recreation, and only half for agriculture.

Enclosure of the proposed Markerwaard polder has begun. The dykes out to Marker Island and the length of the Oostvaardersdiep are completed. The Oostvaardersdiep is to permit smaller ocean traffic to move back and forth from Amsterdam to the North Sea through the locks of the Afsluitdijk. Heavier ocean traffic would continue to use the IJ and the modern giant locks at IJmuiden. With the completion of this work the future expansion to the North and Northeast of Amsterdam would be assured along with greater flexibility of the city's shipping.

However, uncertainty has returned. Planners have discovered that fresh water needs may outweigh those for new land, and perhaps the impolderization of the Markerwaard may not occur. There are five alternative plans before them, but authorities have decided upon none as yet. The securing of Amsterdam's future expansion is not the main advantage of the Zuiderzee project. More important is that such efforts effect the salvation of all of the Netherlands. Planning, cooperation and government control have created a virgin environment which to some measure promises to redress the imbalances existing in the old.

The Delta project illustrates further the advantages to be derived

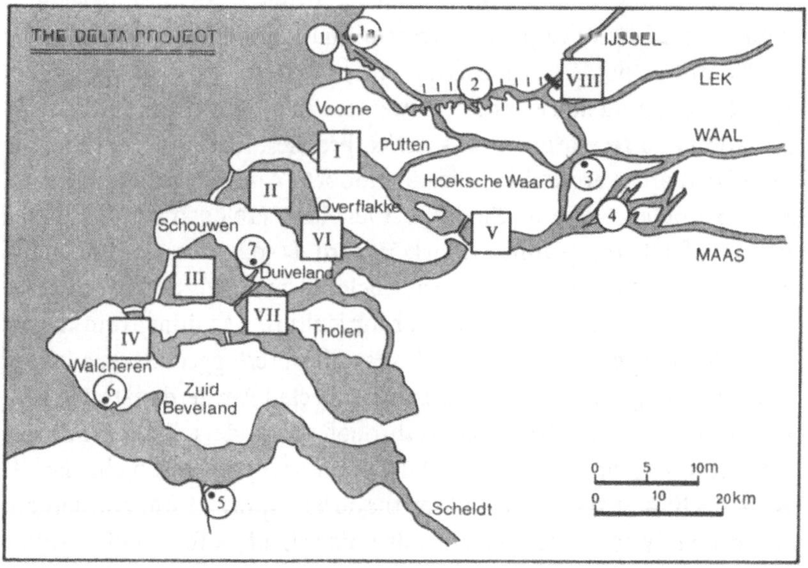

I. Haringvliet	V. Zandkreek	1. Europoort	4. The Biesbos
II. Brouwershavensche Gat	VI. Grevelingen	1a. Hook of Holland	5. Terneuzen
III. Oosterschelde Gat	VII. Volkerak	2. Rotterdam Harbor Area	6. Flushing
IV. Veersche Gat	VIII. Krimpen a/d IJssel	3. Dordrecht	7. Zierikzee

from national planning and public cooperation in the creation and control of environment. A glance at an early map of the Netherlands will make clear that there existed two salient sea arms penetrating the lowlands. One descended from the North into the Zuiderzee aimed at Amsterdam and Utrecht. At one time tidewater did reach Utrecht. Amsterdam in fact arose from the sea. The construction of the Afsluitdijk checked this threat of the sea from the North. Before the Delta project a second arm came from the West, circumventing the Zeeland islands, heading up the Lek, Waal and Maas, threatening Dordrecht. In the disastrous St. Elizabeth flood of 1421 much of this area was inundated. This is the Biesbos (woods of the reeds), lost to the sea only recently to be reclaimed. The Delta Project was conceived to counter the threat from the west. This none too soon, for while plans were maturing, winds, tides and high seas struck in the disastrous February 1953 floods that inundated most of the Zeeland islands and inland areas adjacent to them. In all 375,000 acres, or approximately five percent of the land area of Holland were flooded. Eighteen hundred lives were lost and over half a billion property loss incurred. Yet most material losses were restored within the year. A visitor to the islands today would be hard put to find any evidence of the disaster. The event prompted government decision and mounted public support for the Delta Project.

One must consult a map of southwest Holland and the Zeeland islands in order to comprehend the logic behind the Delta Plan. Four "Afsluiters" (enclosing dykes) are to join the five islands Voorne-Putten, Goeree-Overflakkee, Schouwen-Duiveland, Noord-Beveland, and Walcheren-Zuid Beveland in a defensive front against the sea. The four "Afsluiters" planned were the Haringvliet, Brouwershavensche Gat, Oosterschelde, and the Veersche Gat. All but the Oosterschelde are completed. Its possible disturbance of the region's ecology and attempts to save the local oyster and mussel fisheries formed the basis of opposition to the original proposal. New plans are under discussion for Oosterschelde's future.

Behind these will be a series of three secondary dams vital to the support of the first in that they control the dangerous movements of the waters behind them. These are the Zandkreek, Grevelingen, and the Volkerak. A special flood control gate (Stormvloedkering) at Krimpen on the IJssel to regulate waterflow into the Delta was the first construction of the project. In addition there are a number of backup weirs to control the flow of the Rhine.

The result of the Delta Project will be to turn all but two of the former sea arms into freshwater lakes. The Maas with Europoort at its mouth, and the southern Schelde, the outlet for Antwerp, will be left open for ocean vessel trade. Roads are to be built atop the dykes, thus bringing the former isolated and underpopulated Zeeland in contact with the densely populated Randstad, principally Rotterdam, which like Amsterdam, is also facing strangulation. Having encountered the charm of these islands, one is of two minds about the ultimate results of such progress.

Work was begun in the fifties on all phases of the Delta Plan. Difficult problems have been overcome in the most ingenious of methods. For example, the sluices of the Haringvliet created a special problem in construction. If the supporting dykes were built first the water currents would be too strong to permit construction of the sluices or locks. If the sluices were built first there would be difficulty closing the last gaps of the dykes. Dutch engineers simply built a circular dyke, created an island and constructed the sluices. They were left open while the supporting dykes were attached to them. To help in the speeding up of land fill of the larger dykes the engineers employed telphers, large overhead conveyers that shuttled back and forth from one island to the other. Work has already been completed on many of the projects. The flood barrier at Krimpen aan de IJssel was finished October 1958. The smaller Zandkreek Dam was built October 1960. The larger Veersche Gat, finished April 1961, attached Noord-Beveland to the Walcheren-Zuid-Beveland Island. Grevelingen was finished in 1965 somewhat behind schedule. The three-way bridge-dam at Volkerak was finished ahead of schedule in 1957. Two larger projects, Haringvliet and Brouwershavensche Gat, were completed on schedule in 1969 and 1970.

The benefits of the completed Delta Project will be multiple. Foremost will be the security offered against the sea. Future floods in this region will undoubtedly be controlled. Important also will be the reversal of the salination of the land, particularly in the valuable horticultural area of The Hague – Hook of Holland – Rotterdam triangle. The large freshwater lakes made from the former sea arms, and the water control system created, will force fresh water to course through the Rotterdam waterway reversing the salination of that region. New accesses to the populous areas north of the islands will improve the economic welfare of the farmers. It will also hasten industrialization in this traditionally underemployed region. The lakes and their beaches will be a welcomed recreation playground for all of Holland. The northern islands, particularly Voorne-Putten, will be able to absorb Rotterdam's population overflow.

The IJsselmeer and Delta Projects were grandly conceived. Few American government undertakings exhibit such scope or daring. Considering the complexities and difficulties of the task given it, Dutch marine engineering proved rather impressive. Equally impressive was the willingness of the Dutch to underwrite the rather large construction costs of these projects. In these intelligent attempts to improve their environment the Dutch displayed both imagination and discipline in working towards socially desired goals.

CHAPTER VII

ENVIRONMENTAL PROTECTION

Serious awareness of the threat of pollution came to the Netherlands well after the midpoint of the 20th century. The affluence that creates the economy and problems of waste is fairly new for the Netherlands. In the past refuse disposal was accomplished by building dump heaps or relying upon the ever convenient canals. Then such carelessness seemed to be without problem. However, Holland's wealth and technology are now reaching the "junk" producing levels that America has had for decades. The development of new paper, plastic and synthetic products, coupled with new packaging techniques threatens to inundate her with debris. Garbage production in the Netherlands runs about 4.6 tons, or about 350 pounds per person annually. While this may not seem inordinate when compared with the 815 pounds per capita level of America, it must be realized that the Netherlands is one of the world's most intensively industrialized nations. The problem of waste disposal can only increase. A total of 14 million tons of garbage is predicted for the year 2000. This is a three-fold increase in thirty years, and even then does not include sewage, agricultural wastes, much industrial waste, nor abandoned automobiles. Environmental pollution is an achievement Holland now shares with her fellow industrialized societies.

Garbage and waste disposal in the Netherlands remains a local responsibility. Many communities continue to use the refuse dump, and indications are that this practice will grow – at least doubling by 2000. Much garbage will be used as fill as is presently done in other cities as New York or Tokyo. Because of Holland's high water levels this may not prove to be so practical, particularly in the "quaking bog" regions where fill would be most needed. The Hague has successfully used garbage for the creation of artificial dunes, which quickly become overgrown and hardly distinguishable from the natural environment.

Industrial plants are often required to dispose of their own wastes, much of which is sold as salvage. Many larger cities have arranged to sell or give away the decomposable solid waste which makes up a large bulk of municipal garbage. It was found to be a valuable compost, particularly for the sandy and gritty soils of the eastern Netherlands. The Hague has a long standing contract with Drente (1932–68) whereby its garbage was sent by specially designed railcars to be composted for farm use. In contrast to the Los Angeles proposal to transfer its garbage by rail to dump heaps in Death Valley, the Dutch solution seems more productive. That so few people will ever visit Death Valley will hardly make its desecration less ugly. With the Dutch compost disposal method refuse was beneficially used, and more important "disappeared." It is regrettable that the expenses involved are forcing a decline in the use of this method of refuse disposal.

Much of the new debris of our technological progress seems useless and indestructible as well. Attempts are still being made to maintain composting. There are fifteen mechanical composting plants still in operation, including those of Eindhoven and Groningen. Research to further its use is being carried out by special organizations, foremost being Stichting Compost of the Ministry of Agriculture and Fisheries. It works in cooperation with VAM (Vuilafvoer Maatschappij), an organization founded in 1929 to research economic and scientific methods for disposal of garbage and refuse, to which most of the larger Dutch cities now belong. Undoubtedly the larger cities of Holland will turn more to incineration to solve these problems. The Hague now incinerates its garbage. Incineration is costly and probably will increase local air pollution. Yet estimates are for a twelvefold increase by 2000. Compost disposal will probably decline, and dumping will continue. The result will be to the further pollution of air, water and soil – and more ugliness. The health threat will not only affect man, but animal life and agriculture as well. Clearly the products of progress are not unmitigated desiderata.

Unfortunately there is increasing air pollution in present-day Holland, particularly in the Rijnmond and IJmond regions. There are other local pockets of air pollution, noticeably in the former mining areas of Limburg. A growing source of pollution is the increase of auto traffic. To an American accustomed to exhausts prevalent in New York, Philadelphia or Los Angeles, this seems negligible. One might encounter traffic odors in the bigger cities, but seldom a traffic smog. The use of oil in central heating is another minor source of pollution. However the conversion to the sulfur- free natural gas from the newly developed gas fields of the Northeast has begun the lessening of this obnoxious and dangerous pollution.

The government has attended to this problem by passage of the Public Nuisance Act of 1952, which requires licencing of any construction that might be a pollution source. Enforcement is carried out by cooperation between the government Labor Inspectorate (1890) and the Inspectorate of Environment Hygiene (1962) which can curtail industrial activity on grounds of possible injury to the worker or surroundings. One of the provisions of the license regulations results in the forced construction of higher chimney stacks for potential pollutant sources such as industry, energy plants or incinerators. A simple solution, perhaps applicable to America as well.

Air pollution is difficult to solve on a purely local basis. In the Netherlands regional authorities have been created to attack this problem and more are in formation. First of its kind was the supramunicipal area the Rijnmond created in 1964 by special parliamentary enabling legislation. It encompasses 23 communities in the industrial Rotterdam area – Schiedam, Vlaardingen, Pernis, Botlek, etc. Thus there is the possibility of a coordinated monitoring and control of air pollution created by the complex of oil refineries and chemical plants of the region. This government effort was preceded by that of the Stichting Europoort Botlek-Belangen, an association of industries which, accepting the principle of self-policing, carried on their own research and controls. Dutch industry has been much more responsible in this direction than American or European industry in general. The Netherlands government which is the largest stockholder in the blast furnaces at Hoogovens at IJmuiden has its own staff for air pollution research and control. It has similar services for the mines and new industries in Limburg. Amsterdam, which has developed new chemical and oil industries, such as the Mobil Oil Refinery, also maintains its own research and control staff. Further research is carried out on control and abatement methods of air pollution by the Public Health Engineering Division of TNO. There are, in addition, the "watchdog" efforts of the Rijksplanologische Dienst over Holland's social and physical resources.

The Dutch seem surprisingly alarmed over a problem, which to one accustomed to American industrial dereliction, seems rather minuscule. In the Rijnmond area, despite a three-fold increase in industrial growth in a five year period, air pollution levels remained rather constant. Weather is Holland's ally. When there is ample wind air pollution is acceptable. However, in becalmed warm days pollution sometimes becomes so heavy that factories are obliged to curtail or cease production entirely. A monitoring alarm system automatically alerts authorities to these conditions.

Another contributor to the pollution, congestion and social cost in the Netherlands is the growing ownership and use of that dubious blessing of 20th century technology – the automobile. Like their American counterparts the Dutch have fallen in love with the auto. Just in ten years (1958–1968) auto traffic in the Netherlands rose by 400 percent and is mounting exponentially! There were barely 100,000 vehicles registered before the occupation and not until 1950 was that level exceeded. By 1955 there were one-quarter of a million private vehicles registered, by 1960 over one-half million, by 1965 over 1.2 million, and by 1972 over 3.4 million. Estimates for 1980, 1990 and 2000 are four million, 5.8 million and 7.5 million private registrations respectively! Today only 20 percent of the Dutch families are without cars while 35 percent of the families have two or more.[1] Just where these future autos will be able to park and how they will be able to circulate is a problem that is challenging the ingenuity of the planners on all levels of government. Earlier there were few suggesting the curtailing of its expanding use, but the oil crisis and inflationary costs for auto and gasoline have caused some reappraisal of its advantage as a form of transportation.

Fortunately for the Hollanders, who must often drive in constricted town lanes and narrow dyke roads, most of their vehicles are the small European models. The American type cars comprise 5 percent of registrations and are invariably used by business executives or government officials. The Mercedes-Benz is the prestige car of Holland's rich, and government ministers.

A complicating problem in Dutch car ownership is that it is highest in the more affluent and crowded urban aras. The provinces of North and South Holland and Utrecht have higher auto registration than the more open country of Drente or Limburg. One notable exception to this pattern is the high auto registration among the rich farmers of Oostelijk Flevoland, whose rural isolation force them to use autos.

[1] *Gebiedsindeling en Personenauto Prognose*, *Interim Report No. II*, Stichting Het Nederlands Economisch Instituut, Rotterdam, May, 1969.

The increased use of the auto is causing the decline in motorcycles and "bromfietsen" (motored bicycles) used in recent years. At the time of the occupation cycles had declined to under 50,000 in the entire country. By 1960 a peak of 175,000 was reached. But with affluence even the young could afford autos and cycle registration dropped to within 140,000 in five years. In recent years there has been a significant increase in bicycle sales. The Netherlands has a highly developed and rather costly network of paths built especially for cycle and motorbike traffic throughout the entire country. Distances within the country are trelaively short and the cycle was an excellent mode of transportation. It can still remain a sport and a recreation.

More serious is the decline in the use of public transportation. Since 1963 both rail and bus transport had been on the decline and were decreasing at accelerating rates. From 1963 to 1968 railroad passenger kilometers dropped from 7.911 to 7.355. However passenger kilometers were back to 8,039 by 1972. The Dutch rail system is government owned and operated. It is modern, electrified and efficient. Inconvenient scheduling, delays, outmoded rolling stock and the frequency of accidents of our free enterprise railways seem not to exist. Dutch rail passenger service is clean, comfortable, convenient and speedy. The Dutch would hardly tolerate the service our suburban commuters submit to daily on large city lines such as those of Long Island or Philadelphia.

Still the Dutchman seems to prefer the auto. Certainly, for circuitous or non-scheduled transport it is superior. But it would be a social and economic tragedy if modern and efficient railway systems of the country were to fall into passenger disuse just when the oil crisis and the negative aspects of the auto are making mass transportation so necessary. Bus and tram transportation are also beginning to create problems for municipal authorities, for it is they who are responsible for the operation and profits of inter- and intra-city public transports, and they are watching the process with uneasiness. However, the trend is to promote use of public transport.

Also attendant upon the auto's increasing popularity is the rising accident index levels. While the total of auto accidents seems to roughly parallel increased auto use, there was up to 1973 a disproportionate increase in auto accident fatalities. In 1963 auto fatalities stood at 2007 and by 1971 attained a level of near 3,167.[2] While this is a death

[2] *Statistisch Zakboek*, Table 39. For 1974 a lower rate is certainly to be expected.

rate of five points per 100,000 inhabitants below America's roughly 58,000 auto deaths, it is fifteen points higher per registered vehicle than in the United States.[3] The excessive speeds both in center city and on major highways were probably the major explanation. The autowegen permitted speeds over 70 miles per hour. This has been reduced to 65 miles per hour. The highways are generally well engineered but traffic is extremely heavy. In the cities there is apt to be too much cross traffic and the outmoded "priorité à droite" is a most dangerous rule of the road. It is particularly noticeable in traffic circles where as a consequence traffic comes to a halt rather than being expedited.

Government plans, until recently, operated on the assumption of linear projections of continued road building and car use. The 1970 Ministry of Traffic and Water Control report, *TP 2000*, though dealing with issues such as the enlargement of port and water facilities to further integrate Dutch traffic with that of Northwest Europe, centers upon facilitating auto traffic by building of highways, parking space, traffic control and traffic safety. The report admitted unrestricted use of the private auto is no longer possible. Still it made plans for a continued auto-centered civilization that elicited as early as 1972 heavy blasts from critics.[4]

[3] Americans' driving superiority can be attested to by the following statistics. As can be seen the German is the most reckless – speeds on the autobahnen have no limits and 90 to 100 miles per hour require faultless driving. Italian and French drivers come next, especially true in the countryside. France is now undergoing a rigorous drive to curtail auto fatalities much to the annoyance of the individualistic French motorist. Norway's low rate is understandable. Speeds of 40 to 50 miles per hour are de rigueur. Japan, late comer to the auto age, and wide open Australia also stand high.

Netherlands	71.5 per 100,000 auto registrations
USA	54.4 ,, ,, ,, ,,
Norway	53.4 ,, ,, ,, ,,
New Zealand	56.4 ,, ,, ,, ,,
United Kingdom	59.1 ,, ,, ,, ,,
Denmark	61.2 ,, ,, ,, ,,
France	67.9 ,, ,, ,, ,,
Australia	78.7 ,, ,, ,, ,,
Italy	80.2 ,, ,, ,, ,,
Japan	85.4 ,, ,, ,, ,,
Federal Republic of Germany	126.0 ,, ,, ,, ,,

Source: National Air Pollution Control Administration, 1971.

[4] G. Hupkes and R. Dufour, "Traffic Towards the Year 2000," *Planning and Development in the Netherlands*, Netherlands Universities Foundation for International Cooperation, Assen 1972.

The report was scored for its optimism – perhaps a common myopia of planners. Its projections only go to the year 2000. What will occur beyond that is simply ignored. Also challenged are its bland assumptions that everyone will voluntarily agree to the traffic controls (staggered work hours, weekends, holidays, etc.) and that electronic auto guidance systems will work flawlessly. Costs in the report were vastly underestimated according to the critics. Construction costs given by the critics ran from two to five times those stated in the report. Accident levels were also accused of being underestimated. Cost incidental to accidents would run about 30 billion dollars. These and road building costs would become prohibitive.

More serious is the charge that increased accommodation to the auto will modify irretrievably the physical environment of the country. Cities will be bisected, destroying historic centers, and unspoiled natural areas such as the Veluwe and "Greenbelt" of the Randstad will be chopped into segments. This would be particularly tragic, if as the critics suggest, the auto may soon phase itself out as a viable method of transport in so compact an area as the Netherlands. One study, given present linear statistics on car ownership and car use, predicts that before the year 2000, during the weekends from May through September, auto traffic would be forced to come to a standstill.

Leaving the present Arab oil shortages aside, it is obvious that the present pattern of auto use simply cannot continue. Perhaps it will be the lot of the Netherlands to convince others that the auto is obsolete as an intelligent mode of transportation. Even when oil flows are restored all will have been alerted to the finality of the inevitable exhaustion of this energy source. It would be most foolhardy to continue to be so totally dependent for one's economic well-being upon so hazardous a resource. Undoubtedly this will alter life styles but conceivably that is inevitable. In the meantime one must consider the relatively permanent alterations of the environment made in obeisance to what may prove ot be a most temporary artifact. It must ultimately penetrate our understanding that more is not always better.

The Netherlands' difficulties with water do not end with the battering of her seas, the damage of her floods or the ceaseless erosion of her land. Ironically she has a critical need for this resource, as her supply of pure water is being dangerously overtaxed in recent years. Simultaneously there is the problem of the accumulation of polluted water. While there is water enough it is of the wrong kind.

The agency responsible for providing suitable and adequate water is the Government Institute for Water Supply (Rijksinstituut voor Drinkwatervoorziening, RID), and the agency charged with the control of water pollution is the Government Institute for Sewage Purification (Rijksinstituut voor Zuivering van Afvalwater, RIZA). Though detail operations are conducted by communities, municipalities and water control boards, the overall control at the national level remains with these agencies. This permits more uniform planning and control than if all efforts were just piecemeal. When the 1901 Housing Law was passed it provided that all housing would have available piped water and the Public Health Officers would be responsible for its purity. A national water commission was established in 1910 to study the growing problem. In 1913 it formed the Government Institute for Water Supply in order to have an organization to deal with this critical resource on a continuing basis.

In the past, ground water, even canals and rivers provided the country's drinking water. Larger cities were eventually forced to use adjacent water sources, Amsterdam bringing it in by canal boats from the river Vecht. As population grew such primitive means became inadequate. In 1853 Amsterdam built its first pumping stations and towers, using nearby dune groundwater as its source. Today it has to use as well the waters of the Loosdrecht lakes near Hilversum. By 1874 the Hague established a similar water supply system in her nearby dunes. The same year Rotterdam built an extensive system that also supplies the surrounding towns of Delft, Schiedam and Vlaardingen. Its source was filtered river water, which, due to high pollution levels, is now creating grave problems. Leiden built her systems in 1878, Utrecht in 1883. By 1900 all towns of 10,000 population or more, had their own systems. Before World War II, 688 out of 1056 communities had piped water for their residents. Today, one can say all of Holland is supplied.

The effects on public health have been as one would anticipate. Cholera and typhoid have virtually disappeared. Yet as late as 1903 typhoid incidence ran 55.6 per thousand.[5] Beginning around 1935 many communities in endemic goitre regions placed iodine traces in their water supply. Today iodine is distributed by means of a flour additive in bread baking. After experiments on the value of fluoridation

[5] *Holland and its Drinking Water*, 40th Anniversary Rijksinstituut, The Hague, 1953.

to curtail caries, the Ministry of Social Affairs and Public Health recommended the practice. Today the water supply of half the population is treated. A program of public health education is being carried out to increase public acceptance of fluoridation.

Provisions for water supply is basically a municipal function. All large cities have fulfilled this responsibility and even service nearby communities. Smaller communities normally enter into joint ventures or "corporations." There remain some vestigial private water supply companies but their total output is small. Many industries make provisions for their water needs. The State and Provinces help with technical advice and planning, financial aid (usually low cost loans or payment of interest on loans), quality control of water, and adjudicate over water disputes.

Holland's water problem lies in the future. When authorities first concerned themselves with water planning her population was a bare three million. While today's supplies seem adequate, forecasts of over sixteen million population by 2000 pose difficulties. Per capita consumption has grown and predictions all point to further increases. Moreover industry is developing by leaps and bounds and its demands are exorbitant. Anticipated turn of the century demand will be threefold the present consumption level of approximately 1360 million cubic meters.[6]

The irony of Holland's geography is that though she is almost submerged in water, she will not have enough suitable to her needs. While much of her water pollution is man-made, the genuine problem rests in her geology. Running from west to east, her land is characterized by fine sands of the dunes, peat bogs in the polders and behind them, fine sand and clay of the Utrechtse Heuvelrug and the Veluwe, and finally sand, clay and even some limestone in the East. Salt water lies below the western half of the land, turning to brackish as it surfaces near the polders. The fine sands of the dunes are penetrated by rain water, which forms a huge bubble of fresh water which "floats" upon the salt water below it. However if the dune water is too rapidly ex-

[6] Government estimates of water use are in millions of cubic meters:

year	*public*	*industry*	*total*
1980	900	1600	2500
2000	1500	3000	4500

Vijftig Jaar Drinkwater, Rijksinstituut voor Drinkwatervoorziening, The Hague, 1963.

hausted there is vertical, and horizontal, intrusion of the salt water. This also occurs if there is too extensive use of ground water in the East. Moreover heavy use of fresh water in the polderlands increases their salinization as they are most open to salt water "invasion." Thus there is a geologically "safe" level beyond which groundwater cannot be used.

The IJsselmeer and Delta Regions figured in the government's plans for future water supply. IJsselmeer is already contributing to Holland's water needs as well as improving the quality of agricultural land around Zuiderzee that was rapidly deteriorating due to salinization.[7] The future freshwater lakes created from the blocked sea arms by the Delta Project will not be immediately useable. It will take an estimated 100 years before fresh water from the Waal and Maas will flush out the salinity of the bays and the heavily salted impoldered islands. However the increasing inroads of salinity in Southwest Holland will be reversed.

Concerned over the heavy future demands projected by studies conducted in the sixties, the Ministry of Social Affairs and Public Health developed a Master Plan to assure the nation its fresh water needs. After two years the institute for Drinking Water Supply published its preliminary report in 1970. It revealed that two-thirds of the nation's water supply was ground water. Authorities felt that not more than the then consumption level of 1500 million cubic meters of ground water could be taken from the soil. Future calls for at least 4500 million cubic meters simply could not be met. Unfortunately turning to the alternative of surface water presented problems. The Rhine and the Maas that provided the country with its major surface water were heavily polluted. The Rhine is almost unuseable, especially when at low flow level.[8] The Maas, while purer is characterized by frequent periods of low flow and subject to unpredictable use in its course through Belgium.[9]

[7] Cattle in the area have shown an annual increased milk production of 28.8 pounds per cow due to improved water soil. (*Holland and Its Drinking Water*, 40th Anniversary Rijksinstituut, The Hague, 1953).

[8] See chart page 131.

[9] The Netherlands' heavy dependence upon these precarious sources is seen in these statistics on her water supply:

Rhine	70×10^9	cubic meters per year
Maas	8×10^9	,, ,, ,, ,,
Other Rivers	2×10^9	,, ,, ,, ,,
Rainfall	10×10^9	,, ,, ,, ,,

(Master Plan, p. 4).

The Master Plan suggested infiltration of river water into existing and future dune water systems, and the creation of large water reservoirs for future use. Already the Amsterdam and Hague water supply systems employ the method of "stocking" dune ponds with Rhine river water. Fortunately, water if left intact, undergoes a biological (and partially chemical) self-purification process. The Master Plan called for the building of dune reservoirs in the Veluwe region. This proposal is under heavy attack, for the Veluwe is one of the last recreational and nature preserves of the Netherlands. A large reservoir in the Biesbos area is now serving the Rotterdam area, granting respite to her two costly water purification reservoirs constructed in 1966, already working to capacity. Another reservoir is planned in the Delta lake formed south of Goeree-Overflakkee Island for the augmented population and industrial growth anticipated in the Hellevoetsluis area. Other reservoirs in the Maastricht area will provide the best drinking water in the Netherlands. The Dutch propose to dike parts of the IJsselmeer polders – creating a lake within a lake – to insure fresh water supplies for new towns to be built. This is important for the proposed industrial complex of Lelystad. Concern over water resources has been intense enough to alter original designs concerning the Markerwaard polder. Moreover some fear exists that the heavy water traffic on the IJsselmeer, particularly the Oostvaardersdiep, always contains a threat to IJsselmeer's water. One oil tanker accident like the Torrey Canyon would be disaster.

Thinking ahead the Dutch have constructed a desalination plant at Terneuzen. Their experts are cooperating with Israeli scientists in order to avoid costly duplications in research. Other nations seeking Dutch water expertise include Uganda, Australia, Belgium, France, Surinam and the Netherlands Antilles. The Dutch have cooperated with international organizations such as the International Commission for Protection of the Rhine Against Pollution. This has proved somewhat dissappointing to the Dutch government. Pollution continues from the coal mines and chemical industries of the Ruhr and the discharge from the potassium mines near Colmar, France. The Federal Republic of Germany has made considerable inroads upon her pollution in the last decade, largely out of self- interest since the cities of the Rhine Valley rely upon that river for their water supply. France continues offering financial and other arguments for delaying remedial measures.[10]

[10] The French potassium mines are responsible for half the total Rhine pollution.

Holland is prepared to rely upon her own solutions to her water supply. There is a concerted effort on the part of local water authorities to educate the consumer to more careful water use. Voluntary organizations such as the Commission for Information on Water Production (Commissie Voorlichting Waterleidingbedrijven) and the Association for Water Supply Use (Vereniging van Exploitanten van Waterleidingbedrijven) prepare and distribute booklets and make radio and TV appeals to this end. They cooperate freely with numerous interested private organizations.[11] The National Institute for Drinking Water Supply has intensified research on water supply and use. The extent of Dutch commitment to water self-sufficiency is seen in the growth of the National Institute for Drinking Water Supply. Its personnel increased from 53 in 1938 to 124 in 1963 to over 160 today, of which more than 35 are university trained. This is in keeping with the trend of the times – greater costs, greater complexities, and greater vulnerability for society in the event of technological failures.

Sewage and industrial waste disposal was not a serious problem to Holland until after her post-war industrial "explosion." With customary human lack of foresight they merely discharged waste and sewage directly into their rivers and seas. A Medical Inspection Bill passed in 1873 required public health officials to halt such unsanitary disposal methods. An improvement of this bill was passed in 1875, and in 1920 the Government Institute for Sewage Purification (Rijksinstituut voor de Zuivering van Afvalwater, RIZA) was founded. Initially RIZA was under the direction of the then Ministry of Social Affairs and Public Health but was later transferred to the Ministry of the Waterstaat.

fully 130 kilograms per second of the total load of 275 kilograms per second. (*Master Plan*, p. 5) The Dutch have conducted studies on costs of storage of potassium salts near the mines or of hauling them to the sea, but were unable to induce the French to cooperate. The Rijnwater Foundation, an environmentalist group in the Netherlands, is presently filing suit against the Mines de Potasse d'Alsace, implementing a test case action permissible under an unused regulation of the European Community. It is an attempt to achieve through private action what governments seemed unable to do.

[11] Waterworks Association (Vereniging van Waterleidingbedrijven), Association of Netherlands Water Undertaking (Vereniging van Exploitanten van Waterleidingbedrijven in Nederland – VEWIN), Association of Netherlands' Water Engineers and the Institute for Testing of Waterworks Materials (Keuringinstituut Voor Waterleidingartikelen – KIWA) and other government agencies – Research Institute for Public Health Engineering (IG-TNO), Government Institute of Health (RIV), Government Institute for Water Purification (RIZA) and the Technical University at Delft (TNO).

Control of water pollution was implemented by requiring municipalities and industries to have permits for the discharge of liquid wastes, which were granted by local Water Boards (Waterschappen). The changeover was logical since most of the concerns of RIZA would be with these subordinate administrative bodies of the Waterstaat.

Holland's grave problem of water pollution derives from the familiar pattern of population and industry "over-working" a given resource. The low-lying, wet terrain of Holland is not conducive to ready run-off of polluted waters. There is rainfall, the Rhine, the Maas and a few minor streams to "flush" away this pollution. However the Rhine, which supplies 65 percent of the water for this "flushing" action presents more of a problem than a solution. Along with the Ems estuary and the Veenkoloniën, it comprises the three major regions of pollution in the Netherlands. It well deserves its reputation as "the biggest sewer in Europe."

Before industrialization the Rhine was a jewel, a river of pure beauty, a source of pleasure for the inhabitants of its shores. It was even a fisherman's paradise. In 1875 the number of Rhine salmon put up for auction in Holland exceeded 100,000.[12] Today it is so polluted that no commercial fishing is conceivable. In 1880 its salt load at the border town of Lobith was 40 kilograms per second, but by 1965 it registered a count of 332 kilograms per second! The engineers of RIZA set a "safe" load at 225 kilograms per second. Equally important is the oxygen content of any stream. Studies conducted on its pollution levels clearly reveal that the Rhine is "safe" only at high levels of flow.[13] Rhine river traffic constitutes another danger. This is the accidental, or intentional discharge by cleaning of the barges, of oil or chemical substances into the water. Estimates are that 80,000 to 90,000 tons of pollutants are discharged into the Rhine annually. Much of Rhine water pollution is foreign in origin, though more than half Rhine traffic is Dutch. The probabilities of solutions through international cooperation seem unencouraging. Not only is the Rhine not "safe" as a source of drinking water, it cannot even be employed to "flush" away pollution. Thanks to measures taken recently by governments, the pollution level of the Rhine did not increase, at least, in 1974.

[12] *Importance of the River Rhine for the Water-Economy of the Netherlands*, I. J. Hopmans, RIZA publication.

[13] See accompanying chart p. 131.

LEVELS OF OXYGEN CONTENT OF RHINE RIVER AT LOBITH IN RECENT YEARS
("Safe" level of O_2 given as 5 Mg/litre) (Statistics given by RIZA)

Year	*Condition*	*Percent of Time Below 5MG/L Level*	*Lowest Level Recorded*
1953	Dry	29	4.0
1954	Very Wet	17	4.5
1955	Normal	15	3.8
1956	Wet	8	4.2
1957	Normal	35	3.7
1958	Wet	27	3.5
1959	Dry	42	2.6
1960	Normal	30	3.0
1961	Wet	31	1.6
1962	Normal	35	2.8
1963	Very Dry	50	2.6
1964	Dry	53	1.3
1965	Wet	22	3.5
1966	Wet	30	2.4
1967	Dry	37	2.4
1968	Wet	38	1.8
1969	Dry	43	1.7

From the statistics given two conclusions are immediately evident. One is the increasing percent of the time the Rhine falls below the "safe" level as the years go by. The other is the increasingly lower levels that are recorded as the years go by. Except for the intrusion of "wet" years both trends are constant.

The Eems estuary represents another serious pollution problem. Its present sewage content is given as 15 million PE.s, or population equivalents. Sewage and industrial waste are discharged directly into the Eems from neighboring communities. This region has many synthetic industries which have high levels of pollution. Authorities feel that were proper disposal plants erected most of the pollution would disappear. Understandably, industries are reluctant to undertake the necessary costs.

The Veenkoloniën is the region in northeast Holland that is the center of the paperboard industry, and much sugar beet refining. Except for oil refining, these industries require the greatest use of water in processing. Similar to the chemical and synthetic industries, paperboard and carton manufacturing requires 2000 cubic meters of water per worker annually.[14] Discharge from the paper and sugar beet factories is heavily polluted. Moreover it is discharged into canals in a rather low-lying region where water movement is extremely sluggish. The Veenkolonien is one of the most noisome areas of Holland. However, vigorous efforts have been made to slow or diminish the pollution through modern technical installations.

Pollution rests not in these three areas alone. Sewage and industrial pollution is now endemic to all of the Netherlands. At the end of 1969 there were 418 municipal sewage plants covering 7.2 million population equivalents. This was only about one-third of the national needs. Yet the work of RIZA has been tremendous. Without legislative authority, merely by means of discussion and persuasion it has induced countless municipalities to establish costly sewage disposal plants. Between 1935–1965 over 300 treatment plants were erected in the Netherlands.[15] To maintain adequate standards of purification to meet present- day backlogs would involve approximately 1000 million guilders. Another 650 million guilders will be needed for construction up to 1985. These approximations do not include costs underwritten by private industry to solve their own pollution problems, nor do they take into account the increased costs that will grow from what seems to be an inevitable inflationary economy. Yet the present expenditure for clearing water pollution in the Netherlands is roughly the amount expended for candy for the year.

Hopefully there are already shifts of public attitudes. Water pollution controls are being more forcefully exercised and industries and municipalities are increasing the building of sewage disposal systems.

[14] *Leidingwater*, p. 27, Vereniging van Exploitanten van Waterleidingbedrijven, 1965.

[15] An appreciation of the dimensions of the problem of water pollution in the Netherlands can be gained if we consider the following statistics. In 1964 the total amount of sewage and industrial waste was set at 34 million population equivalents or PE. A PE is average waste created per person per year. By 1985 the increase of population and industry will create a minimum additional waste of ten million PE. The Province of Groningen alone, including plans for future sewage pipelines into the open sea from the Veenkoloniën, will soon be disposing of twelve million PE. Direct discharges into the sea from other regions of the Netherlands will be close to another five million PE. There is also a natural purification factor of approx imately 4.5 million PE. for all of Holland. In addition present sewage disposal plants can account for five million PE. Thus by mathematical count – 44 million PE. demand less 26.5 million PE. in facilities – there will remain a need for disposal facilities capable of handling 17.5 million PE. by 1985.

Modern plants are more likely to be of the biological treatment variety than mechanical. Efficiency of these new type plants which operate upon a "digestion" as well as aeration principle, is superior to that of the mechanical agitation type plants which operate at one-third their level. Approximately 60 percent of existing plants is of the biological treatment type. By-products of the biological treatment plants help reduce the cost of operation. Gases formed are used to supply power for operation or are piped back to municipal gas plants. Dry solid matter, now rendered harmless, is often sold – when not given away – to be used as humus by farmers or even the municipal gardeners.

Public control of water pollution is achieved through three measures. First is the necessity for a permit from local Water Boards to discharge sewage or industrial waste into inland or sea waters. This practice gives control over their own water and catchment areas to the local areas involved. Second is the right of local Water Boards to levy assessments against municipalities and industries in proportion to their contributing pollutant levels. Last is the widening control over the planning and design of sewage treatment plants by the Rijksinstituut for Water Purification (RIZA).

As early as 1920 RIZA was charged with scientific and practical research into the quality of surface water in the Netherlands and the means of protection of these waters against pollution, and advising about the question of what protective measures are to be taken.[16] To carry out these tasks the Institute is presently organized into three branches dealing with the technology of sewage treatment, water pollution research, and control legislation. The last deals with matters pertaining to licences to discharge sewage or industrial wastes, and the setting of rates of levy.

The principal office of RIZA is located in Voorburg but plans are afoot to relocate it, possibly somewhere in the eastern part of the country, in keeping with government plans for diversification and leveling of population densities in the Randstad area. RIZA has also an affiliated research station at Sappemeer in the Noordoostpolder. The institute also mainstains a research vessel, "Dr. Kamps," whose crew routinely check upon river pollution levels, particularly on the Rhine. Government expenditures for the Institute were one-half million guilders in

[16] *Short Description of Task and Working Method of RIZA*, Professor Krul, p. 7, RIZA, Voorburg, undated.

1964, which within five years climbed to one million! Budgetary requests for the 1970's anticipate annual increases of over 50 percent. Thus again we see the pattern of growing complexity, increasing technical direction over public affairs, and an awareness of the need to attend to ecological problems. There are elements of hope if the public can sustain its support for ambitious and costly plans to meet backlog and future demands necessary to maintain tolerable levels of water purity.

There has been no subsequent comprehensive national plan issued by the Ministry of Housing and Physical Planning since its 1966 projection for the year 2000. However, piecemeal studies limited to given studies have been issued by various ministries. These seem to indicate a hesitation, a loss of confidence upon the part of the technocratic planners of Netherlands' future. Overall their projections seem more flexible and open. This is evident by the modifications made upon earlier plans. We have seen the government's uncertainty on how best to use the Markerwaard region – as polder or water and recreation reserve. There is outright opposition to plans for the Waddenzee. Apparently that area will remain untouched.

What seems to be coming to the fore is Dutch recognition that national planning alone is not enough. It is felt that present plans must fit Dutch plans into a wider scheme of a land use pattern throughout Northwest Europe. The prosperous economy of the Netherlands owes its existence to that country's geographic centrality to Northwest Europe's industrial and commercial concentrations. This is strikingly apparent now with the enlargement of the European Economic Community following the United Kingdom's entry. As it is central to the Western European economy, just so is the Netherlands central to its ecology. As an entrepôt nation it is rational that the Netherlands' agricultural sector, though still important, take third place to her industrial and service sectors. This is seen in projections to reduce her agricultural component to four or five percent of her working population,[17] (down from 10 percent in 1963) lifting her service employment to 55 percent, while the industrial employment remained at the 40 percent level. This incidentally places Dutch agricultural employment on the same level as the United States'. American service employment is much higher however, as its industrial employment is slightly over half that of the Netherlands. But the trends are obvious.

[17] *Statistisch Zakboek*, p. 63, Table 93.

One of the Netherlands' largest service activities is traffic and transshipment. Dutch ports are most vital to Dutch economy. Planners foresee that somewhere between 1980 and the year 2000 the Rotterdam-Europoort complex will have reached the limits of its expansion potential. It is for this reason that present planning reports emphasize a stepped- up development of the Schelde and Delfzijl-Eemshaven facilities. In fact tentative discussions entertain proposals to include the entire Dutch North Sea region in future planning. Such a charge has already been given the Ministry of Traffic and Water Control.

Conscious that their country is the hub of Northwest Europe, the Dutch have enlarged their plans to include the economic and ecological needs of their neighbors, no doubt hoping in turn that similar consideration might be made to Dutch needs. Already their country is the corridor for international traffic. Netherlands' ports are the center of trans-channel shipping, and the nation is the corridor of movement of goods to and from the industrial areas of the Ruhr, Belgium and Northern France. Traffic demands are growing insatiably. There are suggestions to enlarge and improve upon the Rhine waterway, the world's busiest, in order to prepare it for the expanded European Community commerce. Present railroad facilities in the Netherlands are superb for passenger service, but their declining commercial use is an ill omen.[18] Only 10 percent of commercial traffic is handled by the railroads, with the canals and waterways responsible for close to 45 percent. Most disconcerting is the growing role of vehicular traffic for commercial shipments. This places an insupportable burden upon the road system of the Netherlands. Perhaps we have here the Achilles heel of the Netherlands economic infrastructure.

Proposals of striking boldness have been advanced for the expansion of Dutch air traffic. The airport centers of Schiphol, Rotterdam, Eelde, Zuid Limburg, Eindhoven, and Twente are projected to increase. Schiphol will only be able to handle 200,000 airport movements per year with its present facilities and saturation will occur by the mid 1980's. Looking ahead to the twenty-first century, tentative proposals call for a super airport to handle 70 million passengers per year! If this were not enough, a secondary airport has been proposed for a 40 million yearly passenger load when the first reaches its capacity, per-

[18] *Statistisch Zakboek*, p. 187, Table 24. An approximate 31 million tons decreased to approximately 21.8 million (STRT) 1972.

haps 25 years after its completion. The profitability of KLM, the importance of the air industry to its economy, the centrality of its position in North West Europe, is such that the Dutch wish to assure themselves that they can capitalize upon this resource indefinitely. Obviously there will be fantastic ecological repercussions to these changes.

Another equally important example of Dutch reliance upon the wider European Community is the role of tourism to her economy. The Dutch anticipate a greater influx of visitors as the integration of the E.E.C. goes on apace. Planners are thinking more of developing her seaside resorts, boating marinas, historic centers and recreational areas in ways to enhance their attractiveness to outsiders. They are considering public relations aspects to make foreigners aware of and desirous of using Dutch facilities and accommodations for relaxation and recreation. In 1972 about 3.8 million foreign visitors vacationed in the Netherlands. With the contracting economy the figure for 1974 will possibly be lower by five to ten percent. The Dutch have long since been one of the most public relations conscious people of the world but now even this has become a consideration of integrated national planning. While aspects of tourism and public relations are the concern of other ministries, the planning and development of recreational facilities to fit into the Northwestern European scheme of things has now become a consideration for the National Planning Service. These tentative explorations into European Community-wide planning are examples of Dutch vigor and thoroughness in looking to the future.

We see in all these efforts an attempt to grapple with the overwhelming pressures that heavy population and technological complexity bear upon a physical environment. In specific situations, where the goals are most clear, such as IJsselmeer and the Delta Project, planning has been near total and the results highly satisfactory. In more peripheral areas the approach has been more causal. Success seems to hinge upon the gravity of the problem as perceived by the populace and government authorities – thus, for example, the solution to water pollution has moved further along than those related to treatment of solid waste. One can say, however, that by and large the Dutch have managed. One outstanding failure, unless drastic action is taken, will be in transportation. If the Netherlander continues to invest his money and affection for the automobile as the primary mode of transportation, the result, in so densely a concentrated population, can only be massive and perpetual traffic jams. Skyrocketing costs in road building, insurance (accident rates can only accelerate as more cars take to the road), maintenance and fuel, would seem to make car use prohibitive. Yet if American experience can serve as an omen one cannot become sanguine on this point. Already the American household is spending as much for its auto (or autos) as it is for rent, and seems prepared to continue doing so. To permit the excellent Dutch mass transportation network to dissolve at a time when all authorities agree it is most needed seems too unDutch. A massive program in public relations to devaluate the advantages of auto use seems in order. It will be an uphill struggle for that instrument of modern technology has become so much a part of our being.

Still one may take heart. The Dutch are taking stock of their situation. Planners are no longer as self-assured as in the past. They show some realization of the limitations built into any comprehensive solution for all planning problems. There is greater comprehension of the dependency that Dutch plans have upon the conditions and acts of others. There is evidence that planning is less directed merely towards economic development and growth, and that planning is headed more in the direction of a better quality of life. This trend seems to be a consequence of the more direct, citizen involvement resulting from growing democratization of the process of social decision making. Perhaps in some areas the change may be too late. Possibly the greatest hope comes from the kind of thinking expressed by the former head of the Physical Planning Service, Mr. J. Vink, that National efforts are not towards "planning for a specific point in time, but a process."[19] Not answers but an approach is the solution to the future.

[19] Idenburg, R., *Spotlight on the Future* (Government Publishing Office) The Hague, 1971, p. 69.

CHAPTER VIII

THE ENGINES OF CHANGE

The Netherlands is well organized for the control of her economy, her physical environment and the distribution of her social services. It is not to be expected then that her government would fail to provide similar planning for her social development. These have been late in developing. Economic planning arrived with the creation of the Central Planning Bureau in 1946, physical planning with the Physical Planning Bureau in 1967, and Dutch social services, though they began much earlier, did not become comprehensively unified until the Reorganization Act of 1967. There is as yet no planning for social development on the scale exhibited by these other efforts. There is an inherent logic in this. In all his actions man has overlooked his need for self- development, perhaps because of his instinctual aversion to the thought, learning, effort, and change that self- or social development implies. The Dutch government has been willing to accept in some measure an obligation to call its citizenry to some standards of quality for a better way of life not totally measured in materialistic things. This call for quality is witnessed in the government efforts for social and community development, cultural stimulation, beautification of the environment, recreation, improved education, the advancement of scientific research, and in the support of family life. Efforts in these directions have not been fully coordinated. Almost without exception government programs for social and cultural improvement have been experimental.

Not unexpectedly the work and jurisdiction of many ministries are involved, and while not necessarily leading to conflict, there is confusion and duplication of effort. Most programs for the improvement of education and research development come under the scope of the Ministry of Education and Science. Most schemes for social and community development fall under the Ministry of Culture, Recreation and Social Welfare. It is partly responsible for poor relief and public assistance programs. Six other ministries make contributions toward social development. The Ministry of Traffic and Water Control has community, as well as economic, development plans for its new polder-lands. The Ministry of Agriculture and Fisheries has programs for relocating phased-out fishing villages and surplus farmers. Programs for sheltered employment of the handicapped, vocational training for adults, and for mental health are offered under the auspices of the Ministry of Social Affairs and the Ministry of Public Health. The Ministry of Economic Affairs and the Ministry of Housing and Physical Planning often create community problems by their decisions to initiate regional economic development schemes or the re-orientation of regions to comply with national physical planning goals. The Ministries of Interior and Finance by their very nature are included in social development planning.

The Ministry of Culture, Recreation and Social Welfare, a tri-partite ministry, plays a prominent role in the diversity of programs that the government has advanced for the social and cultural development of its people. The promotion of art, music, theatre and other cultural affairs is the responsibility of its division of Culture. The Dutch government subsidizes many artists, designers, architects and concert groups with grants for study or work. It holds open competitions for statues, monuments and architectural designs for most of the public buildings it erects. This practice is now almost universally copied by provincial and municipal authorities, and private corporations. Parks, public squares, plazas, shopping malls are dotted with art works. The Lijnbaan in reconstructed downtown Rotterdam, and the capital city of The Hague offer many examples of this attention given to public art. Most of her great men and women, and significant historic events, are memorialized in metal or stone somewhere in the Netherlands. Public art forms tend to be rugged, heavy, solid. Some are creative, but most are intriguing and workmanship-like. Well known are "The Dockworker" in Amsterdam's Jordaan section commemorating the dockworker strike protesting the 1941 Nazi abduction of the city's Jewish population. Less known but equally representative are charming bronze figures such as "Lady with a Parasol" (Eline Vere) on the Groot Hertoginnelaan and "The Flaneur" on the Lange Voorhout, both in The Hague. Unlike the crowded and often-overlooked American monuments, those in Holland, as in Europe generally, are in attractive park settings.

The major Dutch concert orchestras are subsidized by the cities, which in turn get funds for this purpose from the national government. The larger cities of Rotterdam, Amsterdam, Utrecht, The Hague, etc. have their own Concertgebouw or cultural centers, and concert programs attract substantial audiences. Some are world renowned. Smaller communities invariably have local musical groups that present popularly received performances in the town squares or parks. Contemporary Dutch art has not yet been able to earn the reputation of its "Golden Age." Since Mondriaan most Dutch work seems experimental and "artsy." The Dutch seem to have respect for their art, and museums are well cared for and attended. Most bigger cities have their theatre companies. There are two Dutch ballet groups and a Dutch opera. Efforts are being made to build a National Opera Theatre in Amsterdam. Subsidies for theatres, orchestras, opera and ballet by national, provincial, municipal and private sources have been substantial.[1]

A second division of the ministry regulates the museums, national parks, national recreation and sports programs, and special youth and continuing educational schemes for adults. The Netherlands is filled with museums and cultural attractions. The Dutch are a metropolitan people, curious, interested and alive to the fascinations found in human and cultural diversities. They read more than their American counterparts and are ardent collectors, hobbyists, sports fans and amateur students of an endless variety of interests.

There is not a small town that does not have a local museum, should the town, particularly if it is historic, not be a museum itself. Every large city has countless museums and the visitor must stay at least a week just to visit all. Towns evoke history. In The Hague the Dutch government continues to use the Binnenhof for Ministry Offices and the Ridderzaal to house the Estates General. Across the way in the Buitenhof is the Rijksmuseum Gevangenpoort which contains instruments of medieval torture, and once served to imprison the brothers De Witt. To the rear of the Binnenhof is the Mauritshuis, a state art museum exhibiting Rembrandts and other earlier paintings. Two historic buildings, the Huis ten Bosch (1645) and the Paleis Voorhout (1750), serve as the Queen's residence when she is in The Hague. The Hague

[1] These were 82.3 million guilders in the 1970–71 season, and 95.1 million guilders for 1971–72, i.e. roughly an increase from $ 33 million to $ 45 million in one year. *Statistisch Zakboek*, p. 88, Table 14.

Gemeentemuseum and the Mesdag museum are especially dedicated to artists of the Hague School. The Gemeentemuseum also displays Mondriaans, other Dutch and internationally known modern painters, musical instruments and handicrafts. There are numerous small musea devoted to special interests such as coins, railroads and costumes of the past. The Hague's cultural centers include the historic Koninklijke Schouwburg, the modernistic Nederlands Congresgebouw, home of the city's symphony orchestra, and the Kurzaal and Opera Theatre at Scheveningen.

Amsterdam is the museum city *par excellence*. It has at least forty. Best known is the Rijksmuseum with its exhibits of Delftware, ancient silverwork, Dutch Golden Century artists, and above all its Rembrandts. The Rembrandthuis on Jodenbreestraat contains his late etchings that some critics find his most interesting work. The City Museum (Stedelijk Museum) has exhibits ranging from Mondriaans to Calder's modernistic mobiles. The new Rijksmuseum Vincent Van Gogh has the most famous collection of this painter. Of interest is the Jewish History Museum housed in the "Waag" building, and the newly opened Scheepvaartmuseum (historic ships) located in a historic building of the old harbor.

Nor are such centers restricted to large cities alone. Smaller Leiden is equally steeped in history. There are the reconstructed 12th century fortifications, "De Burcht," the Pieterskerk, the University with buildings going back to 1575, and the Lakenhal with its well- known picture collection. A visit to more modern times would include the homes of the Pilgrim leaders who guided the settlement of this religious sect in the New World. Leiden has the university botanical gardens, a museum housing one of the most complete Indonesian ethnological collections in existence, other non-western collections, and a small museum of Dutch military history.

There are other exhibits that are *de rigueur* – among them are those at the Kröller-Müllermuseum near Otterloo, the museum at Enkhuizen, the Outdoor museum near Arnhem and the Tulip Festival at Keukenhof. Kröller-Müller has a garden museum of famous sculptures and the largest known collection of Van Goghs. The Zuiderzee Museum at Enkhuizen contains the vessels and tools used in shipping and fishing, as well as the furniture, clothing, and other household articles depicting the former life of Zuiderzee towns of Enkhuizen, Hoorn, Harderwijk, Blokzijl, Staveren and Hindeloopen. There are scenes of a past way of life gone forever now with the construction of the Enclosure Dam and the new polders. An impressive exhibit focuses on the misery the fishing communities suffered during the Great Depression.

The Outdoor Museum (Openluchtmuseum) is a collection of preserved ancient thatched farm dwellings, estates, windmills, and other buildings out of the Netherlands' past. Over 350,000 visit this museum annually. Though very informative, it lacks the integrity of a given era as does our own Williamsburg. Elsewhere the Netherlands does offer something more complete. There one has the opportunity to visit entire villages in a native surrounding. Most often seen however are the over-commercialized Volendam and the former island of Marken. More authentic are the centers of towns such as Veere, Hoorn, Brielle, Gouda or Zwolle.

The Keukenhof is a Dutch institution. Hundreds of thousands converge upon its annual tulip display. Visitors come from all over Europe and fly in from the Western hemisphere. The display and the endless rows of tulips in the surrounding fertile fields are breathtaking.

The Dutch are the most frequent visitors of their museums and cultural offerings. They come by the busloads on tours organized by churches, trade unions or private associations. School children herded by energetic or harassed teachers come in droves to all manner of exhibits. Foreigners come on guided tours from as far away as Spain, Italy, Yugoslavia. Holland's most assiduous visitors are the West Germans. During the summer their influx is a virtual invasion as they swarm into the coast resorts of Katwijk and Scheveningen. There were over seven and one-half million visitors to museums and expositions in 1970, which increased to over eight and one-half million by 1971.[2] The Dutch make every effort to be accommodating, speaking foreign languages, and stocking foreign delicacies in their shops.

Government on all levels and the business community do what they can to encourage tourism. The tourist organization of the Netherlands (Algemene Nederlandse Vereniging voor Vreemdelingenverkeer, VVV on the signposts) does an excellent job in accommodating the tourist trade. Provinces, municipalities and regional areas maintain their own offices of tourism. The Ministry of Culture, Recreation and Social Work provides much of the regulation, maintenance and financial

[2] 7.695 million in 1970, 8.627 million in 1971 to be precise. Statistisch Zakboek, p. 83, Table 2.

support, for this medley of provincial, regional and municipal services.[3]

The State government under the division of Recreation maintains the nation's parks, recreational areas and camping services. There are of course municipal and private services. The nation's largest recreational area is the Veluwe, now being threatened by all manner of land use demands. Even formerly isolated sanctuaries such as the bird refuge on Terschelling have come under pressure with plans to connect the offshore islands with the mainland. The government is planning future outdoor and recreational sites on reclaimed land. It has developed the Veluwemeer center on the Flevoland polder and the newly created Veersemeer center between the former North and South Beveland islands of the Delta project. It is disturbing to see the phasing- out of natural environments. The extent of density pressures and land use demands is seen by the goverment's need to create man-made "nature."

The third division of the Ministry of Culture, Recreation and Social Welfare concerns itself with services for the family, youth, aged, emigrants, migrant groups and community and regional social development tasks. Probably the most innovative activities in social renewal

[3] Tourism, since it has to do with the honest earning of money, comes under the Ministry of Economic Affairs.

The Dutch are assiduous in their public relations efforts to foster tourism. Their public relations efforts go beyond this moreover, to cover almost every possibility that could be advantageous to their interests, or favorable to the enhancement of their national image. There is a spate of Dutch publications; general, cultural, research, trade journals, tourist brochures, etc., in several languages. They wish their country and its offerings to be well- known throughout the world.

In the United States there is an active Press and Cultural Section of the Consulate General which has offices in Washington, D.C., New York, Los Angeles, and naturally, Holland, Michigan. The Dutch have a flair for this kind of touch. For example, they were the primary people to remember the 300th anniversary of the city of New York by scheduling a performance of the Amsterdam Concertgebouw orchestra in Lincoln Center.

Not only does the national government carry on its public relations programs, but so do the major cities and even cooperating touristic regions. While the national government may offer *Digest of the Netherlands*, Amsterdam, for example, has its *Ons Amsterdam* or *Amsterdam Werkt*. These and public offerings such as *Holland Herald* provide a wealth of Neerlandica. If the public relations of the country and city were not enough, regional and commercial efforts also abound. Presently, public relations people of Amsterdam are hopeful of having their city replace Brussels as the center of the enlarged Common Market. Their efforts are quite creative as they tie together such diverse units as the RAI congress and trade exhibition building, Schiphol airport and the Hague Congresgebouw. In this, as in so many other aspects of Dutch life, we see attempts to think into the future and to organize in preparation for it.

are spearheaded by the Commission for Planning and Social Development (Commissie van Maatschappelijke Ontwikkeling). Functioning somewhat as our own TVA (a model much admired in Europe) it proposes programs matching the regional economic development schemes that the government has decided upon for depressed inner city and rural areas. It is not just the building of housing but also the institution of programs to adjust the new occupants to a new way of life. It is not just the bringing of factories and jobs to the rural depressed areas, but the training of former agricultural folk to live in a newly created technological community.

Immediately following World War II the Netherlands had great problems of unemployment and poverty. Under the Reorganization Act of 1948 the Ministry of Economic Affairs advanced programs of capital investment and vocational training to make technological unemployables fit into the nation's growing economy. While largely successful, pockets of poverty persisted. Changing industrial and residential patterns left many people outside the economic rewards and psychological satisfactions of shared community involvement. This did not apply to just economic minorities such as migrant workers, marginal farm hands, canal barge or caravan dwellers, or imported labor. The Northeast, the Southwest and Limburg, suffered prolonged and serious economic difficulties and cultural deprivations. The Commission for Planning and Community Development therefore created a special Interdepartmental Committee to Stimulate Regional Development (Interdepartementale Commissie Stimuleringsgebieden).

The Interdepartmental Committee is focusing upon three targets. First are the industrially depressed areas under development by the Ministry of Economic Affairs. Second are the land consolidations resulting from the Ministry of Agriculture's programs for larger and more economically feasible farms. Last are the wholly new housing complexes resulting from physical planning of new urban redevelopment schemes. The Committee has also helped IJsselmeer and Delta Project authorities in their efforts to create new communities in reclaimed lands. All of these programs have resulted in new living areas that are not yet communities. The task of the Interdepartmental Committee is to provide social, educational and cultural services that knit these new areas into communities, that will help the displaced to accept a new way of life and a new home.

The Committee employed standard social work techniques. It conducted with the cooperation of the new inhabitants, surveys of the assets and needs of the newly planned community. New community "togetherness" was prompted by lectures, discussions, films, school programs and other informational vehicles. Local inhabitants through role playing, social drama sessions, etc., were encouraged to assume social responsibilities previously inconceivable to them. They learned to form their own local bands, sport teams, bring into being their own churches, found their own parks, schools, health centers, sport and recreational programs. This program of social regeneration, somewhat reminiscent of the American TVA or New Deal programs, has been relatively successful.[4]

The Interdepartmental Committee is housed along with the Ministry of Culture, Recreation and Social Work in an attractive high-rise in the Rijswijk sector of The Hague. The Committee's Chief is Dr. G. Hendricks, a well-known pioneer in the field of social development. Its 180 headquarter staff is matched by 150 workers on the Provincial level. Additional municipal and community workers are almost exclusively volunteers. The Committee publishes its own magazine, *Stimuleringsgebieden* and uses other Ministry publications such as *Knipselkrant CRM*, press clippings of news items relevant to social and cultural affairs in the Netherlands, to keep its community program clientele informed of its purposes and progress. The Committee faces some political difficulties. Ordinarily community welfare programs are not given generous budgetary consideration. Municipal authorities, particularly burgomasters, often feel that Committee action impinges upon their freedom in determining town programs. Another difficulty lies with the "green pool" faction in Parliament, conservationist elements opposed to further industrialization of rural areas. Perhaps the thorniest issue is found in the southern Netherlands where Catholic leadership prefers to accept government subsidies and to work out the region's programs without governmental control. The Committee is small in size and budget, but in all probability will grow. Certainly it represents an interesting concept for our times – that a government should be

[4] "Specificatie van Investerings-Subsidies in het Kader van de Sociale Planning in de Stimuleringsgebieden," *Stimuleringsgebieden* 12th year, No. 3/4, February 1969, p. 7. In a typical year (1967–68) such communities received government subsidies that permitted the inhabitants to build under their own direction 20 community centers, 20 regional sports centers, 10 gymnasiums, 4 swimming pools and other projects to a total of 60 in all.

responsible for more than just the material well- being of its people.

Education in the Netherlands has a respectable history. It has been typically Dutch in its class and pragmatic orientation. In the past it was the children of the rich that received preparatory and university education. Children of the poor were trained as domestics in the homes of the rich or as workers on the farms or in the shops or factories of the wealthy. Most education was conducted by religious institutions and was moral in tone and parochial in outlook.

In the 19th century education became compulsory, more secular and up-to-date. Still some class overtones remain. Children of the lower class tend to be given training in elementary skills and vocational areas. Children of the middle class are trained for more exacting tasks. Those that go on to universities tend to form a professional and intellectual elite. Standards remain high but rather rigid. Strongly democratic in their regard for the equal rights of the individual the Dutch still appreciate that this equality does not necessarily spill over to achievement or ability, and thus are apt to feel attempts at "leveling" are foredoomed to failure. Perhaps this reflects their innate Calvinism, or possibly a touch of realism. American's fierce attachment to democratic idealism may have beclouded their view of education's role in society. A century and a half of the most generous support for free, public education has not noticeably equipped American society to deal in any superior manner with its political, economic or social dilemmas.

Two aspects of Dutch education are quite distinct from our own system. The first is that education in the Netherlands is controlled at the national rather than the municipal level. In fact, viewing the Ministry of Education and Science as a State Board of Education, is probably the best way to understand its role. While Dutch municipal school authorities are responsible for day to day operations it is the Ministry of Education that sets standards, devises curricula, establishes teacher certification and allocates funds for the schools. The recent issue arising from State Court rulings that schools are "unequal" due to the different levels of local school financing in American communities would not occur in the Netherlands. The insupportable tax burden pressing down upon local property owners throughout America is avoided since the national government finances education.[5] All schools, private

[5] Government expenditures for all education in 1963 was approximately 3 billion guilders or 6.2% of national income. In 1971 comparable figures were approximate-

and religious as well, receive monies based on pupil ratios. There are uniform programs, standards, training of teachers and financing throughout. Directors and Inspector Generals working out of the Ministry of Education maintain standards and control.

The second aspect of the Dutch school system that differs from our own is the widescale existence of private and religious schools side by side with the public schools. "The principle ... that parents must be given the opportunity of providing their children with the education that conforms to their way of life, their philosophy or the educational method they prefer ..." is fully supported by government funds and other national resources. Preserving this right of parental determination of their children's educational choices resulted in the establishment of a four-fold school system – public, private secular, Protestant and Catholic. This four-fold orientation parallels the pattern of "verzuiling" that has up to recently dominated almost all of Dutch political, social and cultural life.[6] Barely a third of her school population attend public schools. The bulk of the non-public schools is denominational. In 1971 there were 2700 public primary schools while only 166 were private, non-denominational. These are usually based on some specific teaching approach such as the Montessori Method, which is rather popular among middle class Netherlanders. Slightly over 40 percent of all students attends Catholic schools.[7] This corresponds roughly with

ly 10 billion guilders or 8.2% of national income. A spectacular rise can be anticipated for 1974 and 1975. *Statitsisch Zakboek*, p. 80, Table 15.

[6] "Verzuiling" is the division of Dutch society into four major value orientations based upon Catholic, Protestant, humanistic and socialistic traditions. Dutch sociologists saw each orientation (Zuil, i.e. pillar) as a society within a society for most activities of an individual or family – schools, sports, entertainment, mass media, politics – clustered around one of these traditional orientations. A treatment of this now declining aspect of Dutch life is given in Johan Goudsblom's *Dutch Society*, New York (Random House) 1968.

[7] Enrollment figures given by the Central Bureau of Statistics for 1971 are as follows:

	Totals	*Public*	*Protestant*	*Catholic*	*Other*
Primary Students (Nursery included)	1,955,989	534,668	538,457	822,669	160,195
Secondary Students (Vocationals incl.)	1,055,988	215,787	255,740	396,949	187,512*
University Students (higher vocationals excluded)	112,932	81,874	10,405	14,614	6,039

* Figure is distorted by high enrollment in vocational and training type private schools which comprise 122,058 of students in this category.

(Based on data from 1972 Statistical Yearbook of the Netherlands, The Hague, Staatsuitgeverij) pp. 80–81).

the religious composition of Dutch society. The public and secular schools are over-represented since only 22 percent of the population professes to be non-denominational. The national government subsidizes primary and secondary schools and all universities. The State Universities provide educational opportunities for 75 percent of the nation's post- high school students.

The Dutch do not view religious versus public education as a Church-State issue. The resolution of this controversy was achieved early. During the 19th century an alliance of Catholic and Protestant factions fought for the existence of publicly financed religious schools, which was realized in the first decade of this century. Freedom of choice for religious schools is the one issue that had cemented the Protestant-Catholic coalition despite all others. The continued support of church-related schools is not apt to be challenged though the Ministry of Education still sets down standards, requirements, curricula that are equally mandatory upon all schools.

Dutch education has just undergone a sweeping reorganization triggered by the Secondary Education Act of 1963. The Act provided for far- reaching reforms and modernization of Netherlands' education from primary to university levels. The Act aimed at an educationally beneficial flexibility for students to move from one program to another, one level to another, one school to another, depending upon individual needs. It also aimed at meeting the needs of society for it contained provisions for educational careers to cover manpower shortages in industry and new training to keep abreast of technological changes. Budgetary increases and building programs to meet increased enrollments and broadened programs were also provided.

Under the terms of the Act the Ministry of Education and Science was given authority over standards for all Dutch schools. In typically Dutch fashion the Act set for its ideal a one hundred percent certification program for all Dutch school graduates. It reorganized the elementary school program. It provided for multiple tract secondary programs, and set standards of pupil admission to each of these programs. This was done by compulsory examinations and through screening committees composed of local school authorities. While minor regulations and budgetary powers gave the national government much control, local public and private schools have considerable autonomy. Safeguards were written into the Act to protect the traditional patterns of freedom of religious and private schools. Experimentation was encouraged with the passage of a companion "Experimenteerwet." Provisions guiding the assimilation of the older programs into the new were laid out in the Overgangswet of 1967. Small wonder the Dutch called the 1963 Secondary School Act the "Mammoth Law." The bill was a focal point of much public concern.[8]

It was the intent of the Act that innovations for educational improvement would come from the local school boards that were operating under considerable freedom. Annual reports from each school were to be submitted to the Ministry of Education and Science for its approval. It was anticipated that parent and teacher associations, and interest and religious groups in the non-public schools, would be responsible for such plans. Teacher groups were to be professionally involved and not mere trade unions.[9] It was to be educational democracy by fiat. Voluntary cooperation of parent and teacher groups would help smooth innovation in education and sell a rather revolutionary philosophy to profession and public alike. Generally the results of such an arrangement have been, according to government reports, rather satisfactory.[10]

Present educational programs are considerably removed from the former biases of class. Elementary education operates under the Primary Education Act of 1920. Initially a six year program, beginning at age six, it trained all pupils in a common program centering upon the "three R's." The Dutch call this program "Basisonderwijs" – basic education. In the sixties it was extended to eight years with a future goal of ten years. Day school is now mandatory for all youngsters between 14 and 16 but is difficult to realize. In 1955 a Pre-primary Education Act extended this basic education by providing for a two year nursery training period, the "Kleuterschool." While meant to be a common

[8] After considerable debate it received a favorable 100 to 44 vote in the "Tweede Kamer" and a 49 to 16 vote in the Upper House. Normally an issue as professional in content as this would merit little excitement in the Estates General.

[9] Teacher groups cooperated with government representatives to form "pedagogical centers" to research, improve and implement educational advances. Marshalled along the now familiar "verzuiling" pattern they participated with others in the government's National Planning Procedure Committee through such organizations as Catholic School Committee, Union of Protestant Schools, Foundation for the Promotion of Public Education, General Private School Association, the Union of Netherlands Municipalities, etc.

[10] Interim Report HAVO Schools, 1968–1969, First School Plan 1971–72–73 (Eerste Scholenplan) and the Priority Memorandum (Prioriteitennota), Ministry of Education and Science, The Hague, 1967, 1968, 1969.

program for all children, the socio-economic milieu of a given school, the resources of the parents or a religious organization, can create, as in all societies, profound differences. In 1967 an education decree set up a special program (Buitengewoon Onderwijs) for the physically and mentally handicapped that runs through the entire school career of the child. In 1966 a Special Apprenticeship Training Act was also passed to update the various apprenticeship training programs already in existence throughout the Netherlands.

It is in the secondary school programs that we see built-in differences of education patterns and goals. There has been a reorganization of secondary education under the Secondary Education Act of 1963. It represents a rather thorough changeover towards democratic flexibility based as it is on the American model. It does not yet have the unqualified support of either the public or the teaching profession. Old values die hard and the Dutch have always been demanding of their young both in conduct and in school performance. There is growing insouciance towards learning so reflective of our own society, and football heroes, cheerleaders, school hops are becoming more of a preoccupation of Dutch students. Still school is serious business, for Dutch parents and children alike. Perhaps it was too much so in the past.

The secondary school program is divided into four major tracts: Pre-University (Voorbereidend Wetenschappelijk Onderwijs, VWO); General Secondary (Algemeen Voortgezet Onderwijs, AVO); Technical and Vocational (Beroepsonderwijs, BO); and other Secondary Education. With the exception of the last, all of these are in turn broken into subdivisions of their own.

Students leave the primary school usually at eleven or twelve years of age. They all enter the same program for their first year in secondary schools, known as the "Brugklas" – the Bridge Class. The 1969 reorganization act offers longer and broader educational opportunities. The "Brugklas" permits the student and educational authorities a year's postponement in making the difficult decision concerning the student's future educational program. At the end of the "Brugklas" the student, based on his interests and abilities, elects one of the four major tracts. Class overtones play a keen role in these decisions.

If a student elects, and is selected for a Pre-University program, he may pursue it at a Gymnasium, an Atheneum or a Lyceum. All three are six year programs leading to a certificate of admission to the universities. The Gymnasium teaches Latin and Greek, foreign languages and mathematics. In the fourth or fifth year the student must choose between Department Alpha, concentrating on the classics, or Beta, concentrating on a mathematics-science program. The Atheneum omits classics but emphasizes modern languages, history, geography, law and economics. Here too, the student in his fourth or fifth year must opt for Department Alpha, concentrating on economics and social sciences, or Department Beta, the Mathematics-Science tract. The Lycea represent a compromise of Gymnasia and Athenea programs. These are the "select" programs, for they lead to the universities.

Most of the students go into the General Secondary program. These are divided into three levels – lower (Lager Algemeen Voortgezet Onderwijs, LAVO), middle (Middelbaar Algemeen Voortgezet Onderwijs, MAVO) and higher (Hoger Algemeen Voortgezet Onderwijs, HAVO). The LAVO school (the Dutch invariably use the abbreviations) is a two year course leading to lower level vocational and technical training programs. A LAVO graduate also receives a certificate of admission to the Middle School (MAVO). Thus a student is not necessarily locked into a program. The MAVO school offers three year and four year programs. Those completing the three year course usually move on to technical or vocational training programs on the intermediate levels. The four year certificate makes the MAVO graduate eligible for the higher level of technical or vocational training or for fourth year work in the Higher School (HAVO). HAVO offers a five year program. It prepares its graduates for admission to technical and vocational schools of the higher levels. It is possible for HAVO students to move into the Athenea or Lycea under certain circumstances after they have successfully completed two or three years' work. After three years' study under the MAVO program a student may attend a two year special teacher training program preparing one for primary school teaching. There is an abundance of possibilities. Usually it is difficult to move to much higher educational levels except through a series of steps. Rigid as it may seem, there are elements of flexibility.

It is in the technical and vocational training programs that we find the greatest contrast with American secondary education. While we have allowed our vocational high school programs to deteriorate the Dutch have maintained and improved theirs. They offer training in all fields that prepare their young for practical careers in the job market. The Dutch employer does not have to devise his own training programs to fit the secondary school graduate to his task, for normally the graduate is fully prepared after completing his school training. The Dutch react positively towards technical and vocational training for their children. More ready to accept class differences, they do not look upon the vocational curriculum students with the stigma so often felt in America. The success myth is tempered with realism in the Netherlands. Consequently there may not be the same levels of failure and frustration so frequently encountered in our own educational experiences. Offering the variety of courses that they do there is not the high level of "dropouts" among the weaker students that is so much a part of our educational way of life.

The vocational and technical training programs, like the general secondary educational programs, are geared to three levels – lower, middle, high. A student graduating from one of the secondary school levels normally goes on to the corresponding vocational, technical or trade school program. There are at least ten core areas of vocational training. These include training in technical, nautical, domestic service, agriculture, retail trade, administrative, performing arts, social work, teacher training and apprentice programs.

Primary and elementary teacher training is viewed by the Dutch as being technical or vocational in nature. There are programs for the Pre-Primary, Primary, and Secondary teacher. Social Work is also viewed in the same light. The Pre-Primary program (Opleiding tot Kleuterleidster) offers a three year program for teaching, and a four year program for head-teacher. Admissions is granted to the program with a MAVO certificate, or the completion of two years in the HAVO or pre-university schools. Successful completion of the pre-primary teaching program is required for entry for the head-teacher certificate or primary teacher training programs. The more direct avenue is the HAVO certificate. The primary teacher training program (Opleiding tot Onderwijzer) lasts two years for the elementary teacher certificate, or an additional year for the head-teacher. Teacher training on the pre-secondary level in the Netherlands remains somewhat like our older Normal School pattern. High school teachers are prepared by attending special schools where they are trained for the mandatory state examinations. University graduates and graduates of some higher level technical and vocational schools can also qualify as teachers if they receive certificates through state examinations. Plans are afoot to broaden eligibility to teacher training institutions to HAVO graduates. Social work certificates are obtained from specialized schools. A MAVO certificate permits admission to the intermediate level Social-Pedagogic Program (Sociaal-Pedagogisch Onderwijs, SPO). A HAVO certificate permits admission to the higher level of the program. Under certain conditions a higher level SPO graduate can enter some departments of the universities for further study.

Dutch education, especially vocational training, is thoroughly structured. Understandably it requires an energetic administrative oversight to operate at its optimum. It could be flexible to the needs of the individual provided professionals responsible for its operation remain alert. The grouping of students into selected training programs based upon abilities, interests, needs and capacities is sound. If properly done the task of the teacher might be lightened and his or her efforts more fruitful. Conceivably there should be higher levels of achievement and satisfaction on the part of the individual student.

There is certainly not in the Dutch schools the same situation of high chaos we so frequently find in the schools of our major cities. But we are talking about different societies, with different values and different living styles. Our professionals can hardly do other than what our society permits. It is doubtful that American educationalists would be opposed to a vocational training program on the order of that found in the Netherlands. Many are openly questioning the wisdom of the massive flood of high school graduates encouraged to go on to the universities. Along with non-professionals they may well be pondering over the conveyor-belt aspects of what is politely called "social promotion" in our educational system. Serious citizens appear to be increasingly concerned over the outcome of the egalitarian educational goals that have been proclaimed the salvation of our democracy. Dutch parents are beginning to share some of this concern now that their schools have moved closer to the American model.

Dutch university education is held in esteem. Under the Higher Education Act of 1961, the universities were given specific social responsibilities to provide skills and leadership training for the elite of future Netherlands. They embarked upon long-term expansion programs to meet anticipated needs for future research and development in all fields of science. The Act set up machinery for greater autonomy of university administration, for greater control by faculties over teaching curricula and research, and for broader scope of inter-university coordination and development. There are already plans for further reorganisation and democratization of the universities.

University education has seen a steady growth in the Netherlands. In 1900 a bare .8 percent of eighteen year olds entered the universities. By the 1940's this had climbed to 2 percent, by the 1960's to 4 percent, till today it stands at over 5 percent.[11] While nothing like the 50 percent of American high school graduates who enter colleges of one sort or another, it must be born in mind that over twice as many Dutch students go to the technical and vocational schools as go on to the universities. Many of these schools easily stand on an equal footing with some of our four year colleges. Thus comparisons are not easily made. While probably only half as many Dutch high school graduates go on to levels of higher education as do Americans, we must appreciate that higher education, particularly university education in the Netherlands is purposely elitist.

While students' career choices are left to the individual, his parents, his class background, and his guidance programs in the lower grades have almost pre-determined them. Future employment opportunities enter into the decision. The Dutch constantly study educational statistics, employment figures, work needs, and attempt to guide students into programs meeting the most logical projected manpower needs.

There are twelve universities in the Netherlands. The three historic State Universities are Leiden, Groningen, and Utrecht. There are two universities in Amsterdam, the former city University, and the now rather progressive Calvinist Free University in the Buitenveldert region of that city. The Erasmus University at Rotterdam has a faculty in medicine, law, economics and sociology. There is a Catholic University in Nijmegen and another at Tilburg which has concentrations in law, economics and social science. The former State Technical schools, at Delft, Eindhoven and Enschede, and the Agricultural school at Wageningen have been elevated to university status.

Total university enrollment at the turn of the century stood at 4000, went to 40,000 when the Higher Education Act of 1961 was passed and in 1971 was over 112,000.[12] The heaviest enrollment, about 20,000, is in the University at Amsterdam, which in addition to being in the most dense population center, offers the fullest curricula – arts, science,

[11] *Educational Policy and Planning in the Netherlands: Directorate for Scientific Affairs Organization for Economic Cooperation and Development*, Paris, 1967, p. 71, Table 18. Actual university enrollment in 1971 was 5.5%. *Statistisch Zakboek*, p. 81, Table 17.

[12] *Statistisch Zakboek*, 1973, p. 76, Table 9.

theology, law, social science, medicine and dentistry. Utrecht, Leiden and Groningen also have large student bodies. Among them they have faculties of social science, law, economics, theology, and dentistry. Utrecht has a faculty in veterinary science. Nijmegen has offerings in science, social science, theology and medicine for its 10,000 students body. The Agricultural University at Wageningen has passed the 4000 mark in enrollment. There is also a college of Protestant Theology at Kampen. Erasmus University has over 10,000 and Tilburg, 4000 students. The Free University has about 10,000 students divided among programs in arts, sciences, medicine, law, social science, economics and theology.

Delft is the largest technological and engineering research and training center of the Netherlands. It has the nation's most complete scientific laboratory complex. These include wind tunnels, a hydraulics laboratory (in which there are miniature scale models of the IJsselmeer and Delta projects) and an atomic reactor laboratory. Experiments at the atomic laboratory (Kerncentrum) are geared to work in the actual production of atomic energy at the nation's nuclear energy plant at Petten off the North Sea just south of Den Helder. Delft is the nation's center for research programs in pure and applied science. Most degrees conferred are in civil, mechanical, electrical, chemical, applied physics, aeronautical and hydraulic engineering. A lesser number of degrees are granted in pure mathematics, geodetics, metal science and technology, architecture, and naval construction. Delft is world renowned for its research programs in hydraulics and land reclamation.

The technical University of Eindhoven is located in the city that houses one of Europe's largest electronic complexes, the Philips Corporation. Much of the work done is in research directed and applied to electronics. It offers degrees in chemistry, theoretical or applied physics, scientific methods, pure mathematics, and social science. Enschede Institute was created by the 1961 Higher Education Act. It has already grown to a student population of some thousands. It is the only full campus university in the Netherlands. In all other universities the student must provide for his own living accomodations. Enschede's programs center on electrical and chemical engineering.

There has been an exponential growth in government expenditure for both public and private higher education, far outrunning national income growth and ordinary budgetary increases.

"... the average growth of national income since 1950 has been over 9 percent a year; the total government expenditure on education displayed an average growth percentage of over 14.5 percent a year, while government expenditure on university education rose over 21 percent. The development of the latter expenditure displays a certain acceleration of growth."[13]

Paralleling these expanded programs, increased enrollments, and higher expenditures, was a widening of university participation in planning its own affairs. With characteristic Dutch thoroughness the 1961 Higher Education Law explicitly detailed the structure, powers, functions, lines of communication and responsibility of those in higher education. The Crown was represented after the 1971 reorganization by the Ministry of Science Policy; the professionals by the Academic Council. The latter was made up of three representatives from each university, including the Rector Magnificus (in effect, the University Dean) and 10 Crown appointees with advisory powers only. A University Council provides for each University's autonomy. Students and non-professionals are represented. University Council decisions are subject to Crown veto which is rarely exercised. Each University has a working Executive Committee of from three to five members, including the Rector Magnificus, one or two elected members and matching Crown appointees. It prepares and implements University Council decisions, appoints and discharges staff, handles the legal aspects of university affairs and other tasks somewhat similar to our more familiar, over-worked, college president. There is also a forty man University Faculty Council, one-half of which must be teaching faculty, the other non-teaching. Subject to University Council approval it determines curricula, teaching and research activities, establishes qualification, standards and examination procedures, sets faculty rules and recommends appointments, tenure and promotion. It also serves as a grievance body.

The Dutch University system seems to offer much broader autonomy than is to be found in American universities, even those having modern innovations such as faculty senates or student councils. Undoubtedly this greater faculty and university autonomy is a reflection of the his-

[13] "First and Second Progress Memorandum by Professor Posthumus on University Education Objectives, Functions, Structures." *Higher Education and Research in the Netherlands*, Netherlands University Foundation for International Cooperation, Bulletin 4, Volume 14, 1970, p. 12. See also *Statistisch Zakboek*, p. 80, Table 15.

toric pattern of university independence which began as early as the granting of medieval charters. The extensive democratization of Dutch higher education is beginning to disturb the traditional calm of her universities. More numerous students with rising demands for greater financial assistance, coupled with the disappointments of increasingly unemployed graduates of some disciplines, have widened the gap between students and the scientific and administrative university staffs.

The government efforts for the development and promotion of science are closely allied to the work of the universities. The Netherlands has two semi-government bodies for this purpose; the Organization for the Advancement of Pure Science (Zuiver Wetenschappelijk Onderzoek, ZWO) and the Organization for Applied Scientific Research (Toegepast Natuurwetenschappelijk Onderzoek, TNO). The ZWO and the TNO have their head offices in the same building of The Hague and they work closely together. The Boards controlling the Committee on Pure Research (ZWO) are composed of members from the academic and scientific community. Pure Research is conducted at the various universities and inter-university institutes, and costs are fully underwritten by the state.

The Institute for Applied Research (TNO) came into being primarily through the initiative of the scientific rather than the industrial community. Prominent in gaining government sponsorship for applied research was H. A. Lorentz, Nobel Prize winner in theoretical physics. With the support of the Royal Netherlands Academy of Science he convinced officials to create a government body, the Central Organization for Applied Scientific Research. It was only after World War II that it gained a substantial research and development role in Netherlands' science and industry. In 1934 an Organization for Industrial Research was added; in 1940 an Organization for Food and Nutritional Research; in 1946 a National Defense Research Organization and in 1949 an Organization for Health Research. A reorganization of the TNO is now under discussion, particularly in reference to the financing of its activities.

Presently each of the five TNO divisions is separate and controls its own affairs. They are regulated by non-governmental boards made up of government, scientific and industrial representatives. It is almost impossible to describe the complexity and multivariety of TNO programs.[14] Research is carried on in some of the universities, but more

[14] TNO has conducted research on silicosis in industry, noise levels at airports,

in TNO's own research facilities. The bulk of these are centered in Delft, with overflow installations in Rijswijk and The Hague. Two substantial research facilities are in Utrecht and Leiden, and lesser research projects are conducted in TNO laboratories or stations at Amsterdam, Rotterdam, Groningen, Enschede, Wageningen, Tilburg, Den Helder, IJmuiden, Apeldoorn, Arnhem and Waalwijk. There are some 40 institutions and research units operating under its auspices. TNO has over 5000 employees of whom 800 are university trained engineers and scientists. TNO research is only partially supported by the Netherlands' government. More than half of its income is derived from fees paid by industrial firms or other sources for research TNO has undertaken on a contract basis. Research in nutrition, health and defense is heavily dependent upon government support.

One cannot complete a discussion of technical research in the Netherlands without a reference to the Hydraulic Laboratory at Delft responsible for so much of the nation's progress in marine engineering. It began in 1927 largely through the efforts of the renowned Professor J. Thijsse as a private foundation concerned with problems of water control and land reclamation. The pressing problems at that time were proposals to restore the North Sea Canal, enlargement of the Maas, and the Zuiderzee project. It has made tremendous advancement in applied hydraulic science since. It is now a non-profit organization offering research contracts on a cost-basis fee. Seventy to eighty percent of its contracts is with the government. It has also conducted numerous research projects for foreign governments in the last decade. Frequently it operates at a loss but so important is its research that the national government inevitably covers its obligations.

In its laboratories engineers duplicate on model scale the conditions of wind, wave, water currents and sand movements to study problems construction engineers will encounter with the sea. Models of dykes, dams, locks, sluices, weirs are also designed to test conditions under which they will be finally built to endure. Detailed models of the IJsselmeer and Delta projects were built aiding substantially in their real

artificial lighting in homes and offices, the most efficient means of heat radiation, and air pollution effects upon the façade of San Marco in Venice and dying trees along the boulevards of Paris. Its research on sewage purification came up with an amazingly effective and inexpensive device. Called the "oxidation ditch," it is simply a circular canal around which effluent is pumped for prolonged aeration before being discharged into a receiving river or stream. Innovations like these are numerous.

construction. Presently models of Rijnmond and Europoort are the focus of study at the Delft Laboratory. These are so complete in every working detail that they would be a delight to any boy with a mechanical bent. The Hydraulic Laboratory has a substation on the Noordoost Polder, where among other things it conducts experiments on soil viscosity. It also carries out experiments, for example the effects of ship wave erosion, at the Ship Model Basin at Wageningen. The Delft Laboratory is world known and its research keeps contact and cooperates with that of similarly concerned national and international organizations.

The Dutch direct and develop aspects of their society that Americans could hardly conceive of being subject to control. Social and cultural development, education, and research are concerns directed by the Dutch towards immediate and practical ends. Most alien to us would be their organizational efforts to enhance the quality of life. In our problems with our disadvantaged, minorities, and those dislocated by the misfortunes of life, we are more apt to rely upon money solutions than advice, re-training or close personal concern. We probably view any government directed effort to foster attitudes or alter behaviour habits, regardless of theirwhole someness, as unwarranted intrusions upon individual freedom. Dutch education, while directed towards the fullest possible development of the individual, is conditioned by human limitation and the talent demands of the society. Slotting of individuals into educational levels on the basis of tested potential against the wishes of parent or child would not meet our approval. Still we remain baffled over the high level of student frustrations and the "failure" of our schools. Nor would we feel that career choice should be so dominated by employment market demands. Yet we are often unprepared for the serious imbalances of skills and callings that occur from time to time. Shortages of truly skilled mechanics or medical personnel, periodic gluts of teachers and engineers; overpaid psychiatrists and salesmen, and underpaid musicians and artists become the norm. Failure to foresee the results of cutbacks in our defense and space programs and the slowdown of our family formations and what they meant to our labor market seems indefensible. We are more likely to admire the Dutch programs for immediate and long- range research. Our belief in technological progress would find something "right" in this. We overlook, however, the point that Dutch research is more often paid for, conducted, or controlled by the government – the result being that society at large is the beneficiary. When Dutch private industry uses government research facilities it pays for the privilege often at a profit to the government. While our government makes heavier research investments quite often the profits redound only to private industry.

Our free enterprise system and individualistic orientation make it difficult for us to accept many aspects of the Dutch pattern for individual and social development. We find it difficult to introduce any controls on human choices or opportunities for the individual to move ahead or profit for himself, regardless of what the social consequences may be. Inordinate stress upon the learner or worker, employment dislocations, socially expensive research, gaps in social services, are overlooked in our insistence upon total freedom for the individual. The Dutch are more inclined to feel that social responsibility implies that the individual often needs motivation for his own good.

CHAPTER IX

"... OF ORDER AND METHODS"

Many aspects of Dutch planning and control have been examined to help determine how successful the Dutch have been in structuring their society for economic development, social advancement and life quality improvement. It would appear they have achieved some measure of success from their efforts. They have shown themselves more innovative than Americans in adjusting to many modern problems. It is not to be assumed that their experiences were entirely unique. In the solution of the population, public-medical, employment, housing, economic and ecological problems the Dutch approach fits into the pattern of her neighboring West European, democratic-socialist societies. An assessment of Dutch planning and control could be of possible value, offering suggestions to assist us in a more intelligent ordering of our own American economic, social and environmental affairs. Moreover it would be worthwhile to discover what permits the Dutch easy adjustment to so many modern problems that Americans find difficulty in accepting.

Ecology, the relation of man to his environment, has become the most relevant of the social sciences. Man's present existence is totally encompassed by his social institutions. His adjustment to his environment, thus, becomes largely a problem of these social institutions and the concepts and attitudes that sustain them. How have his institutions and his thinking contributed to ecological dilemmas? What correction of old values and institutions, what inventions of the new, will he make to redress ecological imbalances? For one thing is certain, the imbalances are growing more apparent, and adjustment must be made. Man has the choice of deciding whether that will be by his intelligence or by nature's inevitabilities.

In the past, science seemed to offer humanity unlimited possibilities. It is now becoming apparent that these have invariably been misapplied.

Science now tells us that we can no longer burden the earth with excessive populations or expanding technologies, that continuation of mankind's present course of population and industrial growth could lead to extinction. Serious minds are already pondering the future paths humanity will choose. Will it be through ignorance and failure to act, declining living standards, universal civil strife, global war, or world starvation? Or will humanity be able to alter its understandings and institutions so that a rational program of scientific control of humanity can evolve?

Given the power of modern technology, man's continued adherence to 18th Century Enlightment beliefs of "freedom" and "inevitable progress" would seem to lead towards disaster. American ideology is deeply rooted in this 18th century optimism. The past success of our society ill prepares us for the hard thinking and self-discipline needed to cope with modern problems. Their Calvinist attitudes and a less beneficent geography have kept the Dutch more attuned to economic and ecological realities. Once known as the pragmatists of the world, Americans have become the world's ideologues. We have permitted first our thinking and then our institutions to become out- of- date. The energies of individual enterprise, the unrestricted application of technology, a self-confident morality, and an unbounded faith in our national destiny served to propel our society into worldwide prominence. Little by little our institutions have become stalemated by these beliefs. Our fixation upon notions of Morality, Individual Freedom, The Free Enterprise System, Americanism, Anti-Communism, Motherhood, etc., have made us incapable of dealing with new political, economic, and technological realities.

Few Americans are aware of how little, aside from our economy, we have accomplished since the end of World War II. There has been a minimum of investment of social capital in the society's future. Witness, for example, the erosion of our inner cities, our mass transportation system, our schools and hospitals, our physical environment. While attention to social needs remained minimal for a majority of Americans, the rise of their wealth and living standards were phenomenal.

The US has the world's highest living standards, a condition most visible in our role as private consumers. At the end of the 1960's, we spent $ 2 billion per year on jewelry (as much as the total Federal budget for the War on Poverty); over $ 2.5 billion on foreign travel (matching what the fifty states spend on Welfare); and no less than $ 3 billion on pleasure boating (almost double the government's outlay for foreign aid). More than 60,000 families built swimming pools in the U.S.; and in less than a decade we laid over 10,000 more golf courses at a cost exceeding $ 2 billion (doubling our annual capital outlay for community water supply systems).[1]

Such is our dedication to individualism that we fail to understand that the pursuit of individual goals contributes little to the attainment of desirable social ends.

Even our much vaunted economic development leaves something to be desired. Only recently Northwestern Europe, Japan, and the Soviet Union have exceeded our rates of economic growth. The record of our annual growth rate during the 60's hovered just above 2 percent, while in Northwestern Europe the norm was over 6 percent, with a dazzling 15 percent for Japan! Our 1971 trade balance was the least favorable in our nation's history and continues doing poorly to this time. Our domestic economy has seen severe reversals and our seemingly helpless reliance upon others for energy resources threatens it with further difficulties.

Nor do our social service programs seem any more satisfactory. Our social security payments to the aged are so inadequate that nearly one-fourth of our social security recipients is on welfare. Medical care for our aged and poor leave much to be desired. Housing in our urban communities remains unsatisfactory despite our recent creation of a cabinet post specifically for this national problem. We have had equally disappointing results in solving mass transport problems and the control of our environment. Our level of crime and social disorder is the highest of all advanced nations. While these needs went unheeded during the post-war years, we still found the resources to finance military expenditures that totalled over a thousand billion dollars.

The stresses upon our political and economic institutions are more intense than most of our citizenry would conceive. Responsibility for intelligent management of resources and proper employment of technology can no longer be ignored even by a nation as rich as we reputedly are. There seems little evidence that Americans appreciate how restrictive their historic value patterns have been. Long habituated to moral platitudes and an almost religious faith in outmoded political

[1] Hacker, Andrew, *The End of the American Era*, New York (Atheneum) 1970, pp. 134–135. Copyright 1968, © 1970 by Andrew Hacker.

and economic orthodoxies, many have prevented themselves from developing the open-mindedness that would enable them to understand and to contribute to the intelligent use of the nation's human and other resources for broader creative ends.

The Dutch, on the other hand, have less difficulty in living with the realities of the 20th century. There were in Dutch history forces that formed attitudes similar to our own. Heavily influenced by Calvinism, Dutch society developed a sense of righteousness and a compulsion for success akin to our Puritanism. In many ways Dutch morality is more rigid than our own. The Dutch are notorious for their support of good causes and missionary efforts. Still history and circumstances, the most obvious being the awareness of the relative unimportance of their nation in recent world affairs spared the Dutch from an unconscious association of moral righteousness with that of national greatness. The Dutch exhibit no notions of national grandeur. They feel no self-imposed role to preserve the world from disaster. The eclipse of Dutch power came after the Golden Century and the Dutch have long since accommodated themselves to their place in the world. We remain a world power, albeit a most anxious one, threatened as never before by others whose cultures, styles, and values we feel are so inimical to our own. We have experienced the bitterness of ingratitude of our friends and the panic of loss of world status with the rise of alien political powers. We have driven ourselves to the extremes of irrational anti-communism of which Viet Nam was a glaring example. Dutch nationalism has not taken the turn that creates a phenomenon like our own "Americanism." They do not talk about things "anti-Dutch" or form Committees on Anti-Dutch Activities in their legislatures. There are, of course, individuals with conservative opinions, but they have no Birchites, Minutemen, or White Citizens Councils. The Netherlands, as yet, has no race issue of the dimension existing in the United States. Her fringe groups are more apt to be religiously oriented like the villagers of Staphorst, self-interested like the Boerenpartij, or harmless like the Kabouters of Amsterdam. Not militarily involved in world affairs, they are spared the fantastic military burdens we pay and their taxes go largely towards public benefits. We have yet to learn and fully appreciate the costs of "greatness."

The lesson of Holland's "Golden Century" was not the glory of being an imperialistic world power but the importance of being practical in economic affairs. The Dutch were one of the first to accept modern capitalism and still feel that with appropriate safeguards it is a satisfactory way to produce and distribute their worldly goods. Above all they see capitalism as an instrument rather than as a sacrosanct ideology. The labor and social problems of the 19th century made it clear to the Dutch that unbridled private enterprise had its shortcomings. While using Marxism to diagnose the ills of capitalism, the Dutch declined to employ the Marxist remedy. Instead, they pragmatically chose the middle road of labor and social reform. A tradition of government concern for the economic and social welfare of its citizenry became the Dutch answer. A history of sharing limited resources taught her citizenry to cooperate in husbanding them. In affairs economic, the nation resembles a commonwealth devoted to the welfare of its members. There has been the unspoken cooperation of labor and management for the national economy similar to the pattern of Japan. Dutch corporate structure and labor-management relations are not patterned fully on the competitive, atomistic, and ofttimes chaotic ways of America.

The Dutch have never shared the "embarras de richesses" that is so much a part of the American way of life. The attitudes of "strike it rich" or "the sky's the limit" are not ingrained in the inner consciousness of the Dutch. Theirs is not the American Dream. They have a much lower, and perhaps more realistic, level of expectations. They know they inhabit a crowded, urbanized, industrialized land, and that they are only one among the many competing economies and societies in the world community. They accept the economic controls their government imposes to maintain it as a viable economy for them. Wage cuts or boosts, tax declines or increases, price controls, export-import quotas – these are accepted, not without murmur, but with levels of reason or understanding perhaps more mature than one might anticipate from the American citizen. It is not likely that we would find in America the reasonable cooperation between labor, capital, and government that has undergirded the success of the Dutch program for economic coordination and development since World War II. Nor would we expect his government to help the American understand and accept economic discipline in overcoming the difficulties brought on by the recent inflation and oil payment imbalances.

The Dutch implement their moral concern for their neighbors in a practical and effective manner. Perhaps this is partially because Dutch Calvinism encourages "noblesse oblige," while American Puritanism condemns failure. There is little in our wage, profit or tax structure that would indicate a strong wish on the part of Americans to level out some of the injustices of our economic system. Boasting of ourselves as the richest people in the world, we still tolerate near poverty for a fourth of our population. The Dutch would be appalled at what exists in much of the rural South, our Indian reservations, Appalachia, and our inner cities. In the Netherlands, income distribution is kept more and more in a narrower range. Subsidies for wages in marginal industries, for housing, for the aged, the sick, family allowances – all democratic socialistic measures – are accepted by the Dutch with equanimity. Their Old Age pensions are more fair, rather than being geared to inherently inequitable wage earnings. Preventive public health, and public assistance is their way, rather than our almost cruel system of private responsibility and capitalistic medicine. One area where we are inclined to be more socialistic than the Dutch is in education. This is a ready expression of our optimistic reliance upon the doctrine of "self-help." We are prepared to subsidize educational opportunities of all varieties. The Dutch, perhaps more pessimistic concerning human personality, are less generous. However, by preserving her vocational training tract the Netherlands has provided for herself a pool of secondary skilled and semi-skilled labor so vital to her industrial health.

Another consequence of the smallness of her land and resources has been the readiness of her population long since to accept controls of ecological abuses we are just discovering. The Dutch have never had the luxury of aspirations that accompanies the ownership of "large open spaces." Though a highly industrialized urbanized society of over 200 million, we maintain much of the fiction that we are fresh new land, empty, dotted with farms and Jeffersonian villages. We ignore our slums and our inner cities, flee to the suburbs and dream of the glories of 19th century America. It is far easier for the Dutch to recognize the dangers of speculation and land abuse. Socialist aspects of land control have attained more and more public support. Even the right of the government to be the sole purchaser of any land tracts meets little public resistance. Controls on industry aimed against despoilment of the environment are more easily enforced. Dutch industry would hardly be as influential as American industry in setting levels and timetables on environmental standards. Our auto industry has virtually dictated to the Environmental Protection Agency when exhaust emission standards would be legally mandatory. There would be no need for elaborate and much touted "beautification" programs. The Dutch plan beauty into their highways and other constructions. In the building of the nation's technical infrastructure – roads, harbors, mass transport, public energy, recreational centers, industrial parks – the Dutch are more socialistically inclined. The government has variably found itself a partner with many private enterprises, particularly those which, for one reason or the other, could not operate as fully competitively as its more successful brothers. The Dutch simply have not been afraid to use their economic system, whole or part, for their own social ends of a better life.

Reflecting on the nature of government organization, development, and control of the Dutch society, certain patterns emerge. Perhaps the most obvious is that so little of the issues are as political as they would be in America. Decisions concerning housing, welfare, health and consumer and environmental protection generate nowhere near the emotions or political risks they do in our society. A more casual stance is exhibited by Dutch political leaders. It is as if all more or less acknowledge that the attainment of social goals and the overcoming of social problems must be achieved. One encounters among political leaders, and the people alike, a quiet pride in past accomplishments and a "matter-of-fact, get-the-job-done" attitude.

The effect of this is the creation of a high level of professionalism in the Dutch government concerning control over economic, social, and ecological affairs. There is a general acknowledgement that many problems in these areas lie outside the ken of the public and that solutions are more apt to come from the "expert" than political representatives. Not that the public is indifferent; if anything, the Dutch are more aware than ever of these issues. It is that in general they have put trust in the competenceof their professionals. Political leaders are, therefore, often spared the agonies of decisions and the political risks they entail. The public is also spared the necessity of involving themselves in the intricate details of public policy and management. If the results continue to seem satisfactory the public is content to endow the public authorities with monies and authority necessary to continue their work. With the growing democratization of public interest in government actions there is an increasing level of discussion and criticism of its proposed measures. Still much of what we do with fanfare seems routine in the Netherlands. They have long had unheralded innovations that surpass our much touted HUD's "Latchkey," or HEW's "Headstart" programs. Contributing to this professionalism is the noticeably high career training levels exhibited by Dutch governmental authorities. Aside from the Cabinet posts, few government offices are held by political appointees. The multiple party system of the Netherlands would tone down some of the severities of our two party antagonisms, but it goes beyond that, and for political reasons we tend to keep even technical and professional decisions under political party controls.

A second aspect in the organization of the control, planning and development within the Dutch government is the relatively small size of the organization created. Planning organizations have not as yet fully proliferated into a top-heavy bureaucracy. It is surprising how many of the planning, control and development organizations in the Dutch government are dominated by single individuals. These are largely professionals, highly known before their shift over to government employment. Equally surprising was to discover how small in numbers some of the government agencies are. Most government agencies in the Netherlands are growing in size and functions, but more growth will probably occur in the economic, social and ecological control areas. While this is probably true of other governments, the Netherlands is often pointing the way.

A third, and perhaps the most important aspect of government control and planning, and development, is that invariably, whether agency, committee, or bureau, the body almost always has the effective means or power to do the job of planning, control or development assigned to it. In housing and physical planning agencies, the government has the power to appropriate, with due compensation, whatever land it needs for its nationwide planning schemes. The suggestions of the Central Planning Bureau can be realized since the national government is empowered to control prices, wages, rents, imports and exports. Pollution of air and water can be held in check, for the government has the power to deny discharge permits into rivers and streams, or to assess or fine any violator for excessive pollution. Housing is well controlled also. Permits are required for all building construction, not merely for the routine building code sanctions, but even for architectural style, siting, and future rental rates! There are the normal controls over working safety. The power of the Waterstaat is virtually dictatorial. In all its planning and control efforts the government has effective "bites" that it can impress upon the offender, and Dutch authorities are rigid in the application of the laws. The principle that common social needs take priority over individual property rights is the keystone in the Dutch democratic arch.

A fourth pattern is that government control and planning is not always rational in its organization. Water control is so vital to the Netherlands that one would expect it to be highly centralized, yet, it continues to operate under the older network of waterschappen. The reason for this is purely historic. Like Topsy, the Waterstaat "just grew." There is evidence that a consolidation is underway, but still there is no overall reorganization planned by the authorities. The major explanation for the failure to reorganize is probably just inertia. While the Institute for Drinking Water (RID) comes under the Ministry of Public Health, the Institute for Sewage Purification (RIZA) comes under the Waterstaat. Water research is conducted by the TNO and ZWO both under the Ministry of Education. Some planning for future water use comes under the Ministry of Housing and Physical Planning. A question can be raised as to why Housing and Physical Planning are in tandem in one ministry. The reason seems simply that historically any attempt at planning and control of the environment began with public housing. None of these incongruities are serious.

A final pattern is the manner in which planning, development, and control evolved throughout the society and the government. Historically, the thrust for greater planning, development, and control came from outside the government. Invariably, it was private interest groups that founded societies or organizations that pressured the government into action. First, private citizens felt a social need. Organizations were formed to coalesce private opinion, to inform, and to win over others to a project or line of activity. Once the organization became sufficiently large and powerful, it acted in a semi-governmental role. Eventually, when the government counterpart was created, the founding organization invariably became a key power in the control group set over the new government institution.

The IJsselmeer project was sparked by non-government interest groups. When the government finally undertook the IJsselmeer project, it was Dr. Lely who was invited to oversee the work. The Delta project was conceived and carried out almost exclusively under government direction, but it was forces outside the government that proved to it the necessity and value of ambitious large-scale land reclamation and water control operations. The Central Bureau of Statistics has a parallel history, as does the Central Planning Bureau. There were two attempts at government statistical bureaus before the present Bureau of Statistics was founded in 1899. This came about because the private association, the Union for Statistics, founded two decades earlier, pressured the government into action. University professors of economics had long been urging the government to plan and control its economy. A small, but highly professional interest, it finally prevailed, and the director of the new Planning Bureau became Professor Tinbergen, a renowned authority in the field of econometrics. Similar patterns can readily be seen in the manner in which TNO, ZWO, and the Physical Planning Bureau became official bodies of the Netherlands' government. As public interest involves itself further into ecological and psychological welfare concerns, the government will, in all probability, move in these new directions.

From what has been said, it should be clear that it is not the way things are organized in the Netherlands that get things done. Rather, it is the *willingness* to do them. Due to accidents of history and some deliberate planning, Dutch society is surprisingly well organized for its social and economic development and control. The necessity to cooperate in the struggle against the sea was probably the most sustained historic force behind Dutch acceptance of social over individual priorities. The difficult years of the Industrial Revolution attuned the Dutch to labor controls, worker protection, public housing, and public health. The Depression made social legislation a 20th century necessity. However, it was the German Occupation that stands most immediately responsible for Holland's move into what can be called the "Societal State." The comprehensive planning and control, fostering today's economic and social development in the Netherlands, stem from that time. In view of the successful results of these policies it is unlikely that the Dutch will turn away from their present social system. Possibly the only curtailment of further growth toward broader government responsibility for social and economic development will be experienced when the Dutch citizen feels that further economic and social conrtols infringe upon his core individual rights. These are seen by the Dutch citizen to be largely civil, religious, and intellectual liberties rather than as property rights – though this last must also be respected whenever possible.

Seemingly, the most immediate problem of the Netherlands is her economy. An entrepôt nation, she is at the mercy of world trade. Presently the Dutch economy continues to prosper but is extremely inflationary despite efforts at control. Most recent estimates place the inflationary rate between 8.5 and 9 percent. This is due almost totally to the general inflationary rise in world trade prices. The United States shares considerable responsibility for this international situation with her policy of exportation of her domestic inflation. The Netherlands, in concert with other Western European nations, will be strongly affected by America's recent and future grain commitments to the Soviet Union and the People's Republic of China. So far Netherlands' economy vis-à-vis the United States seems relatively healthy. Still, it is unlikely to assume that the Netherlands will see an expansion of her real GNP in the 70's that will match the record of the 50's and 60's. That past record was largely a reflection of the European Common Market growth, an economic "miracle" not apt to be duplicated. A growth pattern of 45 percent between 1953 and 1960 that occurred in the Common Market can hardly be expected to repeat itself.

Netherlands' overriding concern with world trade is seen in her commitment to Common Market and European policies and her role in all manner of international organizations for peace, economic development, and world stability. She has been steadfast in her full cooperation in EEC matters, even accepting policies unpopular at home. An outstanding example was her implementation of the Mansholt Plan which severely dislocated a part of the Dutch agricultural economy.[2] Nor can the possibility of world depressed markets be dismissed. Troublesome international uncertainties continue and the Netherlands' economic future would seem as precarious as ever.

Today, though marriages are not being postponed, birth rates are declining, initial births delayed, contraception more frequently practiced, abortions more common, as are divorces and extra-marital cohabitation. This is particularly true among the young, where it has the greater social significance. In brief, a revolution in sex and family

[2] The economic aspects of the Mansholt Plan are under attack throughout the Common Market. The Common Agricultural Policy has failed in its promise to lower overall food prices. Its twin programs of protective tariffs on food (aimed at American competition) and price subsidies to the farmer, have kept consumer prices high. Moreover, since the Common Market has little policing powers it has been unable to control frauds that have reached the level of an international scandal. Common Market bureaucrats, fearful that reverses of this sort could only serve to defeat their hopes for a more strongly unified Common Market, are sitting upon the issue. Understandably, the Netherlands' stake in the Common Market is high. See Richard C. Longworth, "Europe's Farm Muddle," *Saturday Review/World*, October 9, 1973.

patterns paralleling the more advanced western nations bids fair to become the fate of the Netherlands. Conceivably, these changes may have come too late. Population estimates envisage 16 to 18 million people in the Netherlands by the year 2000. Traffic experts see a total collapse of vehicular traffic long before the advent of the new century, What disruptions in the economy can be forecast were the auto forced to become *obsolete*? The changing patterns of sex and family are obvious responses to heightened awareness of the demographic dilemma. Undoubtedly the pessimistic Report from the Club of Rome, and the United Nations conferences at Stockholm and Bucharest clearly meant more to the Dutch than they do to the Americans. There have been no recent serious public discussions of the possibility of actually cutting back population levels to enhance the quality of life within the Netherlands. Still, in view of the fact that studies on population and ecology are not summarily dismissed by the Dutch public as Cassandrian wailings, possibly Dutch rationality may even open up to a consideration of this eventuality. The government is heeding this problem and has launched a National Program of Demographic Research.

It is the Netherlands' growing concern for the enhanced quality of life that perhaps will determine the future path of her development. Already the Netherlands is the most "over-used" land. Without her present elaborate technological "infra-structure" and her involvement in world trade, the Netherlands as she now exists would not be viable. With them, and her high degree of organization and industry, the nation enjoys one of the world's highest living standards. Plans call for intensified technological growth to permit greater capitalization upon her advantageous centrality within the heartland of Northwestern European industry. Her future is to accommodate herself to the future growth of her neighbors. The Netherlands is in process of rapidly converting the remainder of her natural environment into a man-made world. Only with the most astute planning, most rigorous controls and best public cooperation can the discomforts of closing space, rising noise levels, sight pollution, social irritants, and mass impersonalization be kept within tolerable levels. Still, continued environmental alteration, like unchecked population growth, must lead inevitably to a point of no return. Will it be the lesson of the Dutch experience that there are human dilemmas beyond resolution?

The Dutch cannot escape the realities of their history, their geography, their economic dependency upon the remainder of the world, the compression of their lives into a limited, highly technological environment. They are a most practical and rational people. They have met the dislocations and restraints of modern life without the intense emotional disturbance that has so beset American society. We have been too emotionally tied up with a deceptively happy past and too frightened to look intelligently into our future. We must learn that many old values and assumptions no longer prevail. Like the Dutch, we must understand that the uninterrupted growth of people and things is neither possible nor desirable. We must learn that we are not alone on this globe and that our welfare is ultimately woven into the fate of others. In all fairness, we must concede that given the circumstances, it is as easy for the Dutchman to be a European and a world citizen as it was for the American to be an isolationist. However, we must learn that even progress has its price. The Dutch are ardent supporters of technology and its blessings but they strive energetically to control its disadvantages. American emphasis upon individualism has lessened our ability to cope with the dehumanization and disappointments of technological civilization. Social disorganization and other dissatisfactions exist, but the Dutch seem more content than Americans despite their wealth and vaunted high standard of living. The Dutch do grumble about not enough room, not enough privacy, not enough freedom, and seemingly this is increasing. But they try to observe the rights of others and their own claim to individual happiness. One gets a distinct impression that through its emphasis upon purposeful work, intimate family ties, and active outdoor living, Dutch society still remains relatively healthy and sane. It is clear that Americans expect too much. Our country is not a cornucopia. We must eventually understand, as the Dutch have perceived, that the future is no longer what it used to be.

To say that the Dutch have conclusively solved the dilemma of civilization's continued growth would be far too laudatory even for one most partial to their virtues. Yet their attempts to plan, order and devise their society merits our appraisal. Perhaps were Americans willing to exercise "... by Force of Order and Methods ..." similar will and discipline, many of their problems would lessen. Conceivably we could learn from their experience that not merely technology or knowledge, but wisdom will determine our future. Dutch experience reveals that the key to their progress goes beyond mere governance to things of the spirit. Their success results less from plans and committees and more from practical common sense, determined will, and energetic effort. Dutch character has been as influential as the Dutch government in creating the "Societal State." While "... Force of Order and Methods ..." bespeak good management, it is not mere technology, but rather vision and determination that will create the future.

GPSR Compliance

The European Union's (EU) General Product Safety Regulation (GPSR) is a set of rules that requires consumer products to be safe and our obligations to ensure this.

If you have any concerns about our products, you can contact us on ProductSafety@springernature.com

In case Publisher is established outside the EU, the EU authorized representative is:

Springer Nature Customer Service Center GmbH
Europaplatz 3
69115 Heidelberg, Germany

Batch number: 10830943

Printed by Printforce, the Netherlands